SIMPLE ARDUINO ONGOING FACE DETECTION AND TRACKING ROBOT, AUDIO PLAYER, BLUETOOTH BIPED BOB (WALKING AND DANCING ROBOT) ETC.., DIY PROJECTS

ACKNOWLEDGMENTS

The writer might want to recognize the diligent work of the article group in assembling this book. He might likewise want to recognize the diligent work of the Raspberry Pi Foundation and the Arduino bunch for assembling items and networks that help to make the Internet of Things increasingly open to the overall population. Yahoo for the democratization of innovation!

INTRODUCTION

The Internet of Things (IOT) is a perplexing idea comprised of numerous PCs and numerous correspondence ways. Some IOT gadgets are associated with the Internet and some are most certainly not. Some IOT gadgets structure swarms that convey among themselves. Some are intended for a solitary reason, while some are increasingly universally useful PCs. This book is intended to demonstrate to you the IOT from the back to front. By structure IOT gadgets, the per user will comprehend the essential ideas and will almost certainly develop utilizing the rudiments to make his or her very own IOT applications. These included ventures will tell the per user the best way to assemble their very own IOT ventures and to develop the models appeared. The significance of Computer Security in IOT gadgets is additionally talked about and different systems for protecting the IOT from unapproved clients or programmers. The most significant takeaway from this book is in structure the tasks yourself.

1.SIMPLE ARDUINO AUDIO PLAYER AS WELL AS AMPLIFIER WITH LM386

Adding sounds or music to our undertaking will consistently make it looks cool as well as sounds significantly more alluring. Particularly in the event that you are utilizing an Arduino and you have loads of pins free, you can without much of a stretch add audio cues to your venture by simply putting resources into an additional SD card module as well as a typical speaker. In this article I will give you that it is so natural to Play music/include audio cues utilizing your Arduino Board. On account of the Arduino people group who have built up certain libraries to construct this in a quick and simple manner. We have additionally utilized IC LM386 here for enhancement and clamor cancelation reason.

Hardware Required:

- Arduino UNO

- SD card

- SD Card Reader module

- 10uf Capacitor (2 Nos)

- LM386 Audio Amplifier

- 1K,10K Resistor

- 100uf Capacitor (2 Nos)

- Breadboard

- Push catches (2 Nos)

- Interfacing Wires

Getting ready with your WAV audio files:

For playing sounds from SD Card utilizing Arduino, we need sound records in .wav design in light of the fact that Arduino Board can play a sound document in a particular configuration that is wav group. To make an arduino mp3 player, there are a ton of mp3 shields are accessible which you can use with arduino. Or in case to play mp3 records in arduino, there are sites which you can be utilized to change any sound document on your PC into that particular WAV record.

So to change over any sound record into wav position, pursue the underneath steps:

Stage 1: Click on "Online Wav Converter" to go into the site.

Stage 2: Arduino can play a wav record in the accompanying configuration. You can toy around with the settings later, however these settings were trial to be the best in quality.

Bit Resolution	8 Bit
Sampling Rate	16000 Hz
Audio Channel	Mono
PCM format	PCM unsigned 8-bit

Stage 3: In the site click on "pick record" and select the document you need to change over. At that point feed in the above settings. When done it should look like this in the beneath picture

Upload your audio you want to convert to WAV:

[Choose File] Daavuya - ...al.com.mp3

Or enter URL of the file you want to convert to WAV:

(e.g. http://cdn.online-convert.com/example-file/audio/m4p/example.m4p)

Or select a file from your cloud storage for a WAV conversion:

♥ Choose from Dropbox   ▲ Choose from Google Drive

Optional settings

Change bit resolution:	8 Bit ▼
Change sampling rate:	16000 Hz ▼
Change audio channels:	mono ▼
Trim audio:	[ ] to [ ]
00:00:00	
Normalize audio:	☐

Show advanced options >

PCM format: [PCM unsigned 8-bit ▼]

[Convert file]   (by clicking you confirm that you understand and agree to our terms)

Stage 4: Now, click on "Convert File" and your Audio document will be converter to .Wav record group. It will likewise be downloaded once the transformation is finished.

Stage 5: Finally group your SD card and spare your .wav sound record into it. Ensure you design it before you include this record. Additionally recall the name of your sound record. So also you can choose any of your four sounds and spare them with names 1, 2, 3 and 4(Names ought not be changed). I have changed over four tunes and have spared them as 1.wav, 2.wav, 3.wav and 4.wav like demonstrated as follows.

	Name	Date modified	Type	Size
⭐ Favorites				
🖥 Desktop	1	23-06-2017 11:37 ...	Wave Sound	3,660 KB
📥 Downloads	2	23-06-2017 07:35 ...	Wave Sound	5,020 KB
📑 Recent places	3	24-06-2017 10:31 ...	Wave Sound	3,301 KB
	4	23-06-2017 07:40 ...	Wave Sound	4,182 KB
☁ OneDrive				
💻 This PC				
🖥 Desktop				
📄 Documents				
📥 Downloads				

Circuit and Hardware:

Circuit Diagram for this Arduino Audio File Player is basic. The total circuit graph is appeared in the Image underneath.

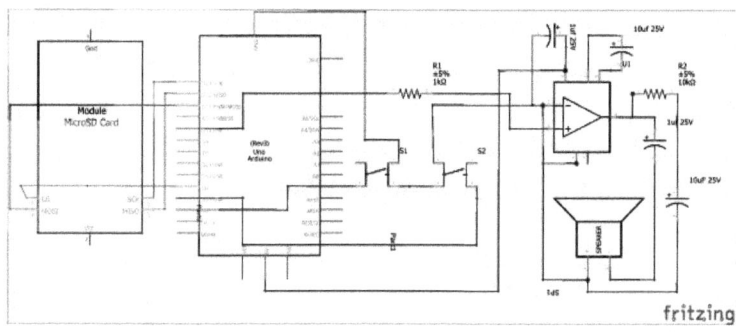

As we probably am aware our sound records are spared into the SD card, subsequently we interface a SD card peruser module with our Arduino. The Arduino and SD card impart utilizing the SPI correspondence convention. Subsequently the Module is interfaced with the SPI pins of the Arduino as appeared above in the graph. It is additionally recorded in the table underneath.

Arduino	SD card module
+5V	Vcc
Gnd	Gnd
Pin 12	MISO (Master In Slave out)
Pin 11	MOSI (Master Out Slave In)
Pin 13	SCK (Synchronous Clock)
Pin 4	CS (Chip Select)

Presently the Arduino will have the option to peruse the music document from the SD card and play it on the stick number 9. In any case, the sound sign delivered by the Arduino on stick 9 won't be discernible much. Henceforth we intensify it by utilizing the LM386 Low voltage Audio speaker IC.

The intensifier appeared above is intended for a Gain of 200 and the Vdd (stick 6) is fueled by the 5V stick of the Arduino. I case you need to build/decline the sound you can expand/decline the voltage gave to this stick. It can withstand a limit of 15V. Become familiar with this 200 increase enhancement arrangement for LM386 here.

We additionally have two push catches associated with the stick 2 and 3 of the Arduino. These switches are utilized to play the following track of the melody and play/stop the music individually. I have utilized these catches just to exhibit its capacities; you can play the melody at whatever point required.

You can collect this circuit totally over a Breadboard as appeared in the image beneath

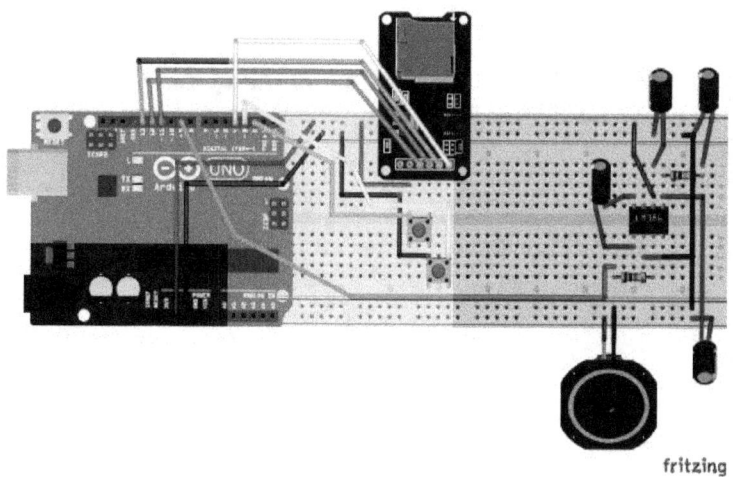

fritzing

Programming your Arduino:

When we are prepared with the Hardware and the SD card, we are only one stage away playing those melodies. Supplement the card into your SD card module as well as pursue the means beneath.

Stage 1: As said prior we will utilize a library to make this venture work. The connection for the library is given beneath. Snap on it and select "Clone or download" as well as pick download as ZIP.

- TMRpcm library

Stage 2: Add this Zip record into your Arduino IDE by choosing Sketch->Include Library - > Add .ZIP Library as appeared underneath and select the ZIP document that we just downloaded.

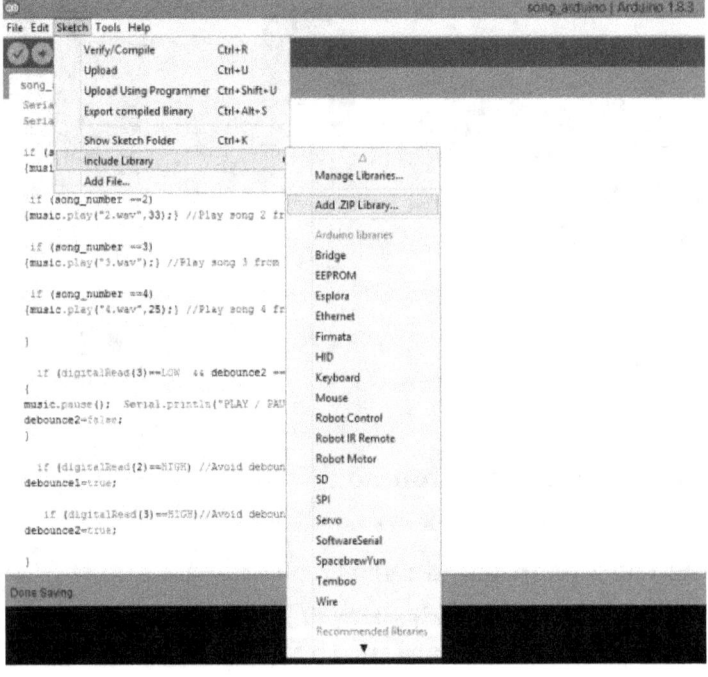

Stage 3: The total program of the arduino music player venture is given toward the finish of this article, basically duplicate it and glue it in the Arduino Program. Presently, click on Upload and prepare to play your sound documents.

The program is clear as crystal since they have the remark lines. Yet, I have likewise clarified the capacity of the TMRpcm library beneath.

Playing a sound document:

You can play any sound that is put away in Wav position inside the SD card module by utilizing the line beneath.

```
music.play("3.wav");

//object name.play ("FileName.wav");
```

You can utilize this line at places where you need to trigger the Audio

Delay a sound File:

To delay an Audio record, you can just call the line beneath.

```
music.pause();

//objectname.pause();
```

Sending/Rewinding an Audio:

There are not immediate approaches to advance or rewind an Audio document, however you can utilize the line underneath to play a tune at a specific time. This can be used to advance/rewind with some extra programming.

```
music.play("2.wav",33); //Plays the song from 33rd second

//objectname.play("Filename.wav",time in second);
```

Setting the nature of the sound:

The library gives us two characteristics to play the music, one is to play as typical mode the other to play with 2X oversampling.

```
music.quality(0); //Normal Mode

music.quality(1); //2X over sampling mode
```

Setting the Volume of the sound:

Indeed, you can control the volume of the sound through programming. You can essentially set the volume by utilizing the line beneath. Higher music volumes will in general influence the nature of the sound, henceforth use equipment control whenever the situation allows.

```
music.setVolume(5);     //Plays the song at volume 5

//objectname.setVolume(Volume level);
```

Working of this Arduino Music Player:

Subsequent to programming your Arduino just press the catch associated with stick 2 and your Arduino will play the main tune (spared as 1.wav) for you. Presently you can press the catch again to change your track to the following tune that is to play 2.wav. In like manner you can explore to every one of the four tunes.

You can likewise play/Pause the melody by squeezing the catch associated with stick 3. Press it once to delay the tune and press it again to play it from where it halted.

Expectation you delighted in the venture. Presently it is dependent upon your innovativeness to utilize them in your undertakings. You can make a talking clock, voice colleague, talking robot, voice ready security framework and considerably more.

Code

```
/*
Arduino Based Music Player
 This example shows how to play three songs from SD
card by pressing a push button
 The circuit:
 * Push Button on pin 2 and 3
 * Audio Out - pin 9
 * SD card attached to SPI bus as follows:
 ** MOSI - pin 11
```

```
** MISO - pin 12
** CLK - pin 13
** CS - pin 4
*/


#include "SD.h" //Lib to read SD card
#include "TMRpcm.h" //Lib to play auido
#include "SPI.h" //SPI lib for SD card
#define SD_ChipSelectPin 4 //Chip select is pin number 4
TMRpcm music; //Lib object is named "music"
int song_number=0;
boolean debounce1=true;
boolean debounce2=true;
boolean play_pause;
void setup(){
music.speakerPin = 9; //Auido out on pin 9
Serial.begin(9600); //Serial Com for debugging
if(!SD.begin(SD_ChipSelectPin)) {
Serial.println("SD fail");
return;
}
pinMode(2, INPUT_PULLUP); //Button 1 with internal pull up to chage track
pinMode(3, INPUT_PULLUP); //Button 2 with internal pull up to play/pause
pinMode(3, INPUT_PULLUP); //Button 2 with internal pull up to fast forward
music.setVolume(5);  // 0 to 7. Set volume level
```

```
music.quality(1);    // Set 1 for 2x oversampling Set 0
for normal
//music.volume(0);    // 1(up) or 0(down) to control
volume
//music.play("filename",30); plays a file starting at 30
seconds into the track
}
void loop()
{


  if(digitalRead(2)==LOW && debounce1 == true) //
Button 1 Pressed
  {
 song_number++;
 if(song_number==5)
 {song_number=1;}
 debounce1=false;
 Serial.println("KEY PRESSED");
 Serial.print("song_number=");
 Serial.println(song_number);
 if(song_number ==1)
  {music.play("1.wav",10);} //Play song 1 from 10th
second
 if(song_number ==2)
  {music.play("2.wav",33);} //Play song 2 from 33rd
second
 if(song_number ==3)
 {music.play("3.wav");} //Play song 3 from start
 if(song_number ==4)
```

```
{music.play("4.wav",25);} //Play song 4 from 25th second
 if (digitalRead(3)==LOW && debounce2 == true) // Button 2 Pressed
 {
 music.pause(); Serial.println("PLAY / PAUSE");
 debounce2=false;
 }
 if (digitalRead(2)==HIGH) //Avoid debounce
 debounce1=true;
 if (digitalRead(3)==HIGH)//Avoid debounce
 debounce2=true;
}
}
```

2. INTERFACING SSD1306 OLED DISPLAY WITH ARDUINO

The majority of us would be comfortable with the 16×2 Dot framework LCD show that is utilized in the vast majority of the tasks to show some data to the client. Be that as it may, these LCD showcases have a great deal of restriction in what they can do. In this instructional exercise we will find out about OLED shows and how to utilize them Arduino. There are loads of kinds of OLED shows accessible in the market and there are heaps of approaches to make them work. In this instructional exercise we will examine about its arrangements and furthermore which will be most appropriate for your task.

Hardware Required:

- 7pin 128×64 OLED show Module (SSD1306)

- Breadboard

- Arduino UNO/Nano

- PC/Laptop

- Associating Wires

Getting to know about OLED Displays:

The term OLED means "Natural Light radiating diode" it utilizes a similar innovation that is utilized in the vast majority of our TVs however has less pixels contrasted with them. It is genuine amusing to have these cool looking presentation modules to be interfaced with the Arduino since it will make our tasks look cool. We have secured a full Article on OLED presentations and its sorts here.

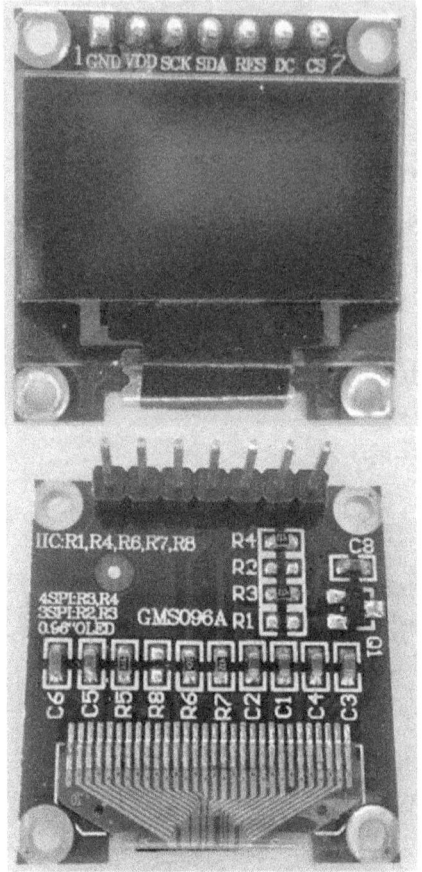

We are utilizing a Monochrome 7-stick SSD1306 0.96" OLED show. The purpose behind picking this presentation is that it can take a shot at three unique correspondences Protocols, for example, the SPI 3 Wire mode, SPI four wire mode and IIC mode. This instructional exercise will cover how to utilize the module in SPI 4-wire mode as it is the quickest method of correspondence and the default one.

The pins and its capacities are clarified in the table beneath.

Pin Number	Pin Name	Other Names	Usage
1	Gnd	Ground	Ground pin of the module
2	Vdd	Vcc, 5V	Power pin (3-5V tolerable)
3	SCK	D0,SCL,CLK	Acts as the clock pin. Used for both I2C and SPI
4	SDA	D1,MOSI	Data pin of the module. Used for both IIC and SPI
5	RES	RST,RESET	Resets the module (useful during SPI)
6	DC	A0	Data Command pin. Used for SPI protocol
7	CS	Chip Select	Useful when more than one module is used under SPI protocol

In this instructional exercise we will basically work the module in 4-Wire SPI mode, we will leave the rest for some other instructional exercise.

Arduino people group has just given us a great deal of Libraries which can be legitimately used to make this much less difficult. I evaluated a couple of libraries and found that the Adafruit_SSD1306 Library was extremely simple to utilize and had a bunch of graphical choices thus we will utilize the equivalent in this instructional exercise. Be that as it may, if your undertaking has a memory/speed requirement have a go at utilizing the U8g Library as it works quicker and involves less program memory.

Hardware and connections:

The Circuit Diagram for SSD1306 OLED interfacing with Arduino is extremely basic and is demonstrated as follows

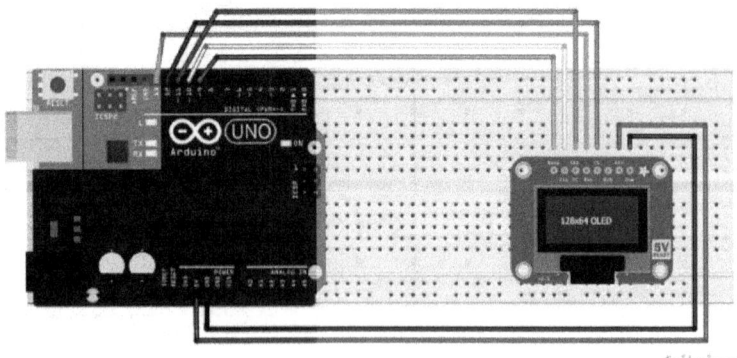

We basically have build up a SPI correspondence

between the OLED module and Arduino. Since the OLED keeps running on 3V-5V and devours almost no power it needn't bother with an outside power supply. You can just utilize wires to make the association or utilize a breadboard as I have utilized so it is anything but difficult to test. The association is additionally recorded in story beneath

S.No	Pin Name on OLED module	Pin Name on Arduino
1	Gnd, Ground	Ground
2	Vdd, Vcc, 5V	5V
3	SCK, D0,SCL,CLK	10
4	SDA, D1,MOSI	9
5	RES, RST,RESET	13
6	DC, A0	11
7	CS, Chip Select	12

Note: You won't have the option to imagine any backdrop illumination/sparkle on the OLED module just by controlling it. You need to program it accurately to see any progressions on the OLED show.

Programming the SSD1306 OLED display for Arduino:

When the associations are prepared you can begin

programming the Arduino. As said before we will utilize the Adafruit library and GFX library for working with this OLED module. Pursue the means to trial your OLED show.

Stage 1: Download the Adafruit Library and the GFX library from Github utilizing the connection underneath

- Adafruit Library

- GFX Graphics Library

Stage 2: You ought to have download two Zip records. Presently add them to your Arduino by following

Sketch->Include Library - > Add Zip library as demonstrated as follows. At that point select the library we just downloaded. You can choose just a single library at once, consequently you need to rehash this progression.

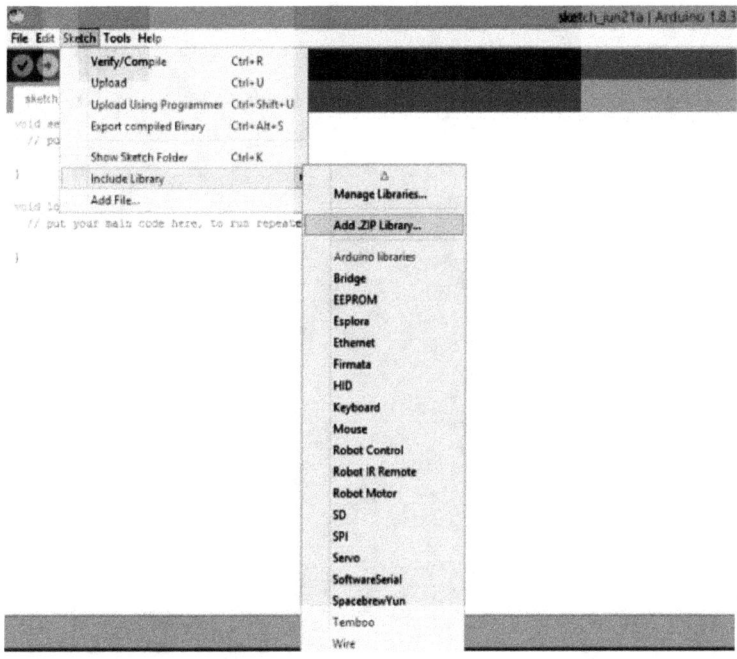

Stage 3: Launch the model Program by choosing File->Examples->Adafruit SSD1306 - > SSD1306_128*64_SPI.ino as appeared in the picture beneath.

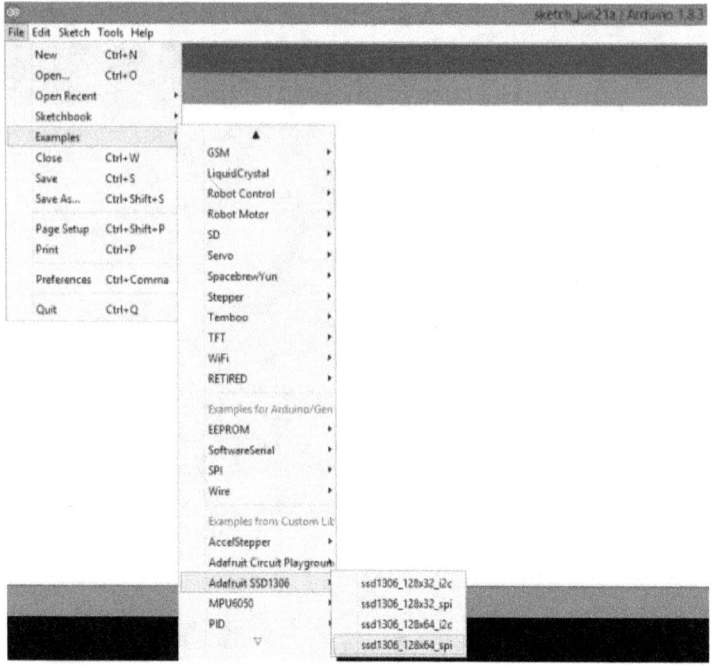

Stage 4: Inside the model program over line 64 include the line "#define SSD1306_LCDHEIGHT 64" as appeared in the picture beneath.

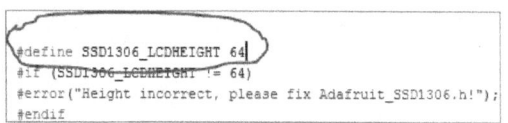

Stage 5: Now transfer the program and you should see the OLED show starting up with the default Adafruit model code as appeared in the beneath picture.

This model program gives all of you potential illustrations that could be shown in the OLED screen. This code ought to be sufficient for you to make bitmaps, draw lines/circles/square shapes, play with pixels, show scorch and string with various textual styles and size and so forth...

In case you have to comprehend the Library and its capacities better you can peruse further. Every garbage of the code is part and clarified with the assistance of remark lines. Complete code is given toward the finish of this Article

Showing and clearing the screen:

Composing on OLED screen is much the same as composing on a chalkboard, we need to compose the qualities and afterward clean it before it could be overwritten. The accompanying directions are utilized to compose and clear the showcase

```
display.display(); //Write to display

display.clearDisplay(); //clear the display
```

Showing a Character Variable:

To show the substance inside a variable the accompanying code can be utilized.

```
char i = 5; //the variable to be displayed

  display.setTextSize(1);   //Select the size of the
text

  display.setTextColor(WHITE);        //for mono-
chrome display only whit is possible

  display.setCursor(0,0); //0,0 is the top left corner
of the OLED screen

  display.write(i); //Write the variable to be dis-
played
```

Drawing a Line,Circle,Rectange,Triangle:

On the off chance that you need add a few images to your presentation you can utilize the accompanying code to draw any of the accompanying

```
display.drawLine(display.width()-1,   0,   i,   dis-
play.height()-1, WHITE);

//void drawLine( x0, y0, x1, y1, color);

display.drawRect(i,   i,   display.width()-2*i,   dis-
play.height()-2*i, WHITE);
```

```
//void drawRect( x0, y0, w, h, color);

display.drawTriangle(display.width()/2,        dis-
play.height()/2-i,display.width()/2-i,disp

lay.height()/2+i,          display.width()/2+i, dis-
play.height()/2+i, WHITE);

//void drawTriangle( x0, y0, x1, y1, x2, y2,
color);

display.drawCircle(display.width()/2,          dis-
play.height()/2, i, WHITE);

//void drawCircle( x0, y0, r, color);
```

Attracting a String to the Screen:

The accompanying piece of code can be utilized o show any message in the screen at a specific spot and size

```
display.setTextSize(2); //set the size of the text

 display.setTextColor(WHITE); //color setting

 display.setCursor(10,0); //The string will start at
10,0 (x,y)
```

```
display.clearDisplay(); //Eraser any previous dis-
play on the screen

display.println("Hello World"); //Print the string
here "Hello world"

display.display(); //send the text to the screen
```

Showing a bitmap picture:

One untrusting thing that should be possible with the OLED module is that it tends to be utilized to show bitmaps. The accompanying code is utilized to show a bitmap picture

```
static    const    unsigned    char    PROGMEM
logo16_glcd_bmp[] =

{ B00000000, B11000000,

  B00000001, B11000000,

  B00000001, B11000000,

  B00000011, B11100000,

  B11110011, B11100000,

  B11111110, B11111000,
```

```
B01111110,B11111111,

B00110011,B10011111,

B00011111,B11111100,

B00001101,B01110000,

B00011011,B10100000,

B00111111,B11100000,

B00111111,B11110000,

B01111100,B11110000,

B01110000,B01110000,

B00000000,B00110000};
```

display.drawBitmap(XPO], YPOS, bitmap, w, h, WHITE);

//void drawBitmap(x, y, *bitmap, w, h, color);

As should be obvious, so as to show a picture the bitmap information must be put away in the program memory in type of PROMGMEM order. Basically, we need to teach the OLED show how to manage every pixel by passing it a grouping or qualities from a n exhibit as appeared previously. This cluster will contain the bitmap information of the picture.

It may sound confused yet with the assistance of a web instrument it is particularly simple to change over a picture into a piece guide esteems and burden them into the above cluster.

Just burden the picture and alter the settings to get your favored see of the picture. At that point click "Produce Code" duplicate the code and glue it into your Array. Transfer the program and you are altogether done. I took a stab at showing a batman logo and this is the way it turned out.

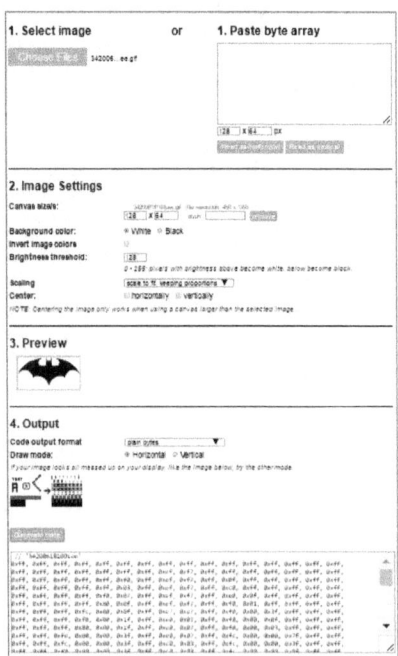

There are still a great deal of things that you can do with these libraries. To realize the total potential outcomes visit the Adafruit GFX illustrations Primitives page.

Expectation you prepared this running and to actualize an OLED show in a portion of your undertakings.

Code

```
/********************************************************************
This is an example for our Monochrome OLEDs based on SSD1306 drivers

 Pick one up today in the adafruit shop!
 ------> http://www.adafruit.com/category/63_98
This example is for a 128x64 size display using SPI to communicate
```

4 or 5 pins are required to interface

Adafruit invests time and resources providing this open source code,

please support Adafruit and open-source hardware by purchasing

products from Adafruit!

Written by Limor Fried/Ladyada for Adafruit Industries.

BSD license, check license.txt for more information

All text above, and the splash screen must be included in any redistribution

```
*********************************************************************/
#include <SPI.h>
#include <Wire.h>
#include <Adafruit_GFX.h>
#include <Adafruit_SSD1306.h>
// If using software SPI (the default case):
#define OLED_MOSI 9
#define OLED_CLK  10
#define OLED_DC   11
#define OLED_CS   12
#define OLED_RESET 13
Adafruit_SSD1306 display(OLED_MOSI, OLED_CLK, OLED_DC, OLED_RESET, OLED_CS);
/* Uncomment this block to use hardware SPI
#define OLED_DC   6
#define OLED_CS   7
#define OLED_RESET 8
Adafruit_SSD1306 display(OLED_DC, OLED_RESET, OLED_CS);
```

```
*/
#define NUMFLAKES 10
#define XPOS 0
#define YPOS 1
#define DELTAY 2

#define LOGO16_GLCD_HEIGHT 16
#define LOGO16_GLCD_WIDTH 16
static const unsigned char PROGMEM
logo16_glcd_bmp[] =
{ B00000000, B11000000,
  B00000001, B11000000,
  B00000001, B11000000,
  B00000011, B11100000,
  B11110011, B11100000,
  B11111110, B11111000,
  B01111110, B11111111,
  B00110011, B10011111,
  B00011111, B11111100,
  B00001101, B01110000,
  B00011011, B10100000,
  B00111111, B11100000,
  B00111111, B11110000,
  B01111100, B11110000,
  B01110000, B01110000,
  B00000000, B00110000 };
#define SSD1306_LCDHEIGHT 64
#if (SSD1306_LCDHEIGHT != 64)
#error("Height incorrect, please fix Adafruit_SS-
D1306.h!");
#endif
```

```
void setup() {
 Serial.begin(9600);
 // by default, we'll generate the high voltage from the
 3.3v line internally! (neat!)
 display.begin(SSD1306_SWITCHCAPVCC);
 // init done


  // Show image buffer on the display hardware.
  // Since the buffer is intialized with an Adafruit
splashscreen
  // internally, this will display the splashscreen.
 display.display();
 delay(2000);

 // Clear the buffer.
 display.clearDisplay();

 // draw a single pixel
 display.drawPixel(10, 10, WHITE);
 // Show the display buffer on the hardware.
  // NOTE: You _must_ call display after making any
drawing commands
 // to make them visible on the display hardware!
 display.display();
 delay(2000);
 display.clearDisplay();

 // draw many lines
 testdrawline();
 display.display();
 delay(2000);
 display.clearDisplay();
```

```
// draw rectangles
testdrawrect();
display.display();
delay(2000);
display.clearDisplay();

// draw multiple rectangles
testfillrect();
display.display();
delay(2000);
display.clearDisplay();

// draw mulitple circles
testdrawcircle();
display.display();
delay(2000);
display.clearDisplay();

// draw a white circle, 10 pixel radius
        display.fillCircle(display.width()/2,    display.height()/2, 10, WHITE);
display.display();
delay(2000);
display.clearDisplay();

testdrawroundrect();
delay(2000);
display.clearDisplay();

testfillroundrect();
delay(2000);
display.clearDisplay();

testdrawtriangle();
delay(2000);
display.clearDisplay();
```

```
 testfilltriangle();
delay(2000);
display.clearDisplay();
// draw the first ~12 characters in the font
testdrawchar();
display.display();
delay(2000);
display.clearDisplay();
// draw scrolling text
testscrolltext();
delay(2000);
display.clearDisplay();
// text display tests
display.setTextSize(1);
display.setTextColor(WHITE);
display.setCursor(0,0);
display.println("Hello, world!");
 display.setTextColor(BLACK, WHITE); // 'inverted'
text
 display.println(3.141592);
 display.setTextSize(2);
 display.setTextColor(WHITE);
    display.print("0x");  display.println(0xDEADBEEF,
HEX);
 display.display();
 delay(2000);
 display.clearDisplay();
 // miniature bitmap display
```

```
display.drawBitmap(30, 16, logo16_glcd_bmp, 16,
16, 1);
display.display();
// invert the display
display.invertDisplay(true);
delay(1000);
display.invertDisplay(false);
delay(1000);
display.clearDisplay();
// draw a bitmap icon and 'animate' movement
                testdrawbitmap(logo16_glcd_bmp,
LOGO16_GLCD_HEIGHT, LOGO16_GLCD_WIDTH);
}
void loop() {
}
void testdrawbitmap(const uint8_t *bitmap, uint8_t
w, uint8_t h) {
uint8_t icons[NUMFLAKES][3];


 // initialize
for (uint8_t f=0; f< NUMFLAKES; f++) {
 icons[f][XPOS] = random(display.width());
 icons[f][YPOS] = 0;
 icons[f][DELTAY] = random(5) + 1;


   Serial.print("x: ");
 Serial.print(icons[f][XPOS], DEC);
 Serial.print(" y: ");
```

```
   Serial.print(icons[f][YPOS], DEC);
   Serial.print(" dy: ");
   Serial.println(icons[f][DELTAY], DEC);
  }
  while (1) {
   // draw each icon
   for (uint8_t f=0; f< NUMFLAKES; f++) {
    display.drawBitmap(icons[f][XPOS], icons[f][YPOS],
bitmap, w, h, WHITE);
   }
   display.display();
   delay(200);


    // then erase it + move it
   for (uint8_t f=0; f< NUMFLAKES; f++) {
    display.drawBitmap(icons[f][XPOS], icons[f][YPOS],
bitmap, w, h, BLACK);
    // move it
    icons[f][YPOS] += icons[f][DELTAY];
    // if its gone, reinit
    if (icons[f][YPOS] > display.height()) {
     icons[f][XPOS] = random(display.width());
     icons[f][YPOS] = 0;
     icons[f][DELTAY] = random(5) + 1;
    }
   }
  }
}
void testdrawchar(void) {
```

```
display.setTextSize(1);
display.setTextColor(WHITE);
display.setCursor(0,0);
for (uint8_t i=0; i < 168; i++) {
 if (i == '\n') continue;
 display.write(i);
 if ((i > 0) && (i % 21 == 0))
  display.println();
}
display.display();
}
void testdrawcircle(void) {
 for (int16_t i=0; i<display.height(); i+=2) {
         display.drawCircle(display.width()/2, dis-
play.height()/2, i, WHITE);
  display.display();
 }
}
void testfillrect(void) {
 uint8_t color = 1;
 for (int16_t i=0; i<display.height()/2; i+=3) {
  // alternate colors
        display.fillRect(i, i, display.width()-i*2, dis-
play.height()-i*2, color%2);
  display.display();
  color++;
 }
}
void testdrawtriangle(void) {
     for (int16_t i=0; i<min(display.width(),dis-
```

```
play.height())/2;i+=5){
        display.drawTriangle(display.width()/2,   dis-
play.height()/2-i,
        display.width()/2-i, display.height()/2+i,
            display.width()/2+i, display.height()/2+i,
WHITE);
  display.display();
 }
}
void testfilltriangle(void){
 uint8_t color = WHITE;
                        for              (int16_t
i=min(display.width(),display.height())/2;i>0;i-=5){
        display.fillTriangle(display.width()/2,   dis-
play.height()/2-i,
        display.width()/2-i, display.height()/2+i,
            display.width()/2+i, display.height()/2+i,
WHITE);
  if(color == WHITE) color = BLACK;
  else color = WHITE;
  display.display();
 }
}
void testdrawroundrect(void){
 for (int16_t i=0; i<display.height()/2-2; i+=2){
  display.drawRoundRect(i, i, display.width()-2*i, dis-
play.height()-2*i, display.height()/4, WHITE);
  display.display();
 }
}
```

```
void testfillroundrect(void) {
 uint8_t color = WHITE;
 for (int16_t i=0; i<display.height()/2-2; i+=2) {
   display.fillRoundRect(i, i, display.width()-2*i, dis-
play.height()-2*i, display.height()/4, color);
  if (color == WHITE) color = BLACK;
  else color = WHITE;
  display.display();
 }
}


void testdrawrect(void) {
 for (int16_t i=0; i<display.height()/2; i+=2) {
     display.drawRect(i, i, display.width()-2*i, dis-
play.height()-2*i, WHITE);
  display.display();
 }
}
void testdrawline() {
 for (int16_t i=0; i<display.width(); i+=4) {
  display.drawLine(0, 0, i, display.height()-1, WHITE);
  display.display();
 }
 for (int16_t i=0; i<display.height(); i+=4) {
  display.drawLine(0, 0, display.width()-1, i, WHITE);
  display.display();
 }
 delay(250);
```

```
  display.clearDisplay();
 for(int16_t i=0;i<display.width();i+=4){
  display.drawLine(0,display.height()-1,i,0,WHITE);
  display.display();
 }
 for(int16_t i=display.height()-1;i>=0;i-=4){
     display.drawLine(0, display.height()-1, display.
width()-1,i,WHITE);
  display.display();
 }
 delay(250);


  display.clearDisplay();
 for(int16_t i=display.width()-1;i>=0;i-=4){
         display.drawLine(display.width()-1,  dis-
play.height()-1,i,0,WHITE);
  display.display();
 }
 for(int16_t i=display.height()-1;i>=0;i-=4){
         display.drawLine(display.width()-1,  dis-
play.height()-1,0,i,WHITE);
  display.display();
 }
 delay(250);
 display.clearDisplay();
 for(int16_t i=0;i<display.height();i+=4){
  display.drawLine(display.width()-1,0,0,i,WHITE);
  display.display();
 }
```

```
for (int16_t i=0; i<display.width(); i+=4) {
     display.drawLine(display.width()-1, 0, i, dis-
play.height()-1, WHITE);
  display.display();
}
 delay(250);
}
void testscrolltext(void) {
 display.setTextSize(2);
 display.setTextColor(WHITE);
 display.setCursor(10,0);
 display.clearDisplay();
 display.println("scroll");
 display.display();


 display.startscrollright(0x00, 0x0F);
 delay(2000);
 display.stopscroll();
 delay(1000);
 display.startscrollleft(0x00, 0x0F);
 delay(2000);
 display.stopscroll();
 delay(1000);
 display.startscrolldiagright(0x00, 0x07);
 delay(2000);
 display.startscrolldiagleft(0x00, 0x07);
 delay(2000);
 display.stopscroll();
}
```

3. ONGOING FACE DETECTION AND TRACKING ROBOT UTILIZING ARDUINO

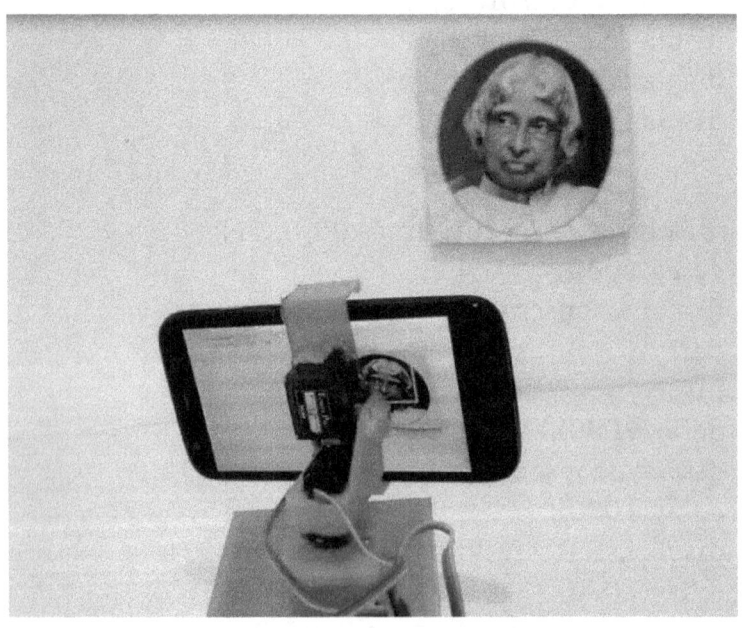

Ever needed to manufacture a Face Tracking Robotic Arm or Robot by basically utilizing Arduino and no other programming like OpenCV, visual fundamentals C# and so on? At that point read along, in this venture we are gonna to execute face location by mixing in the intensity of Arduino and Android. In this undertaking, the portable camera will move alongside your face with the assistance of servos. The benefit of utilizing the Android Mobile Phone here is that you don't have to contribute on a camera module and the entire picture discovery work should be possible in the telephone itself, you needn't bother with your Arduino associated with your PC for this to work. Here we have utilized Bluetooth Module with Arduino to speak with Mobile remotely.

The Android application utilized in this venture was made utilizing Processing Android, you can either legitimately introduce the application by downloading the APK document (read further for connection) or put on your programming top as well as make your own all the more engaging Android Application utilizing the Processing Code given further in the Tutorial. Get familiar with Processing by checking our past Processing Projects.

Before the finish of this instructional exercise you will have a Mini Tilt and Span Robotic Arm that could follow your face and move alongside it. You can utilize this (with further progression) to record your video blog recordings or even take a selfie with the back camera of your cell phone since it positions your face precisely at the focal point of your versatile screen. So!! Sounds fascinating?.. We should perceive how we can manufacture one...

I have attempted my best to make this undertaking to fill in as straightforward as would be prudent, anybody with least learning on equipment or coding can utilize this rules to make this task work instantly. Anyway once you cause it I to recommend you to get behind the codes so you can truly realize what makes this thing work and how.

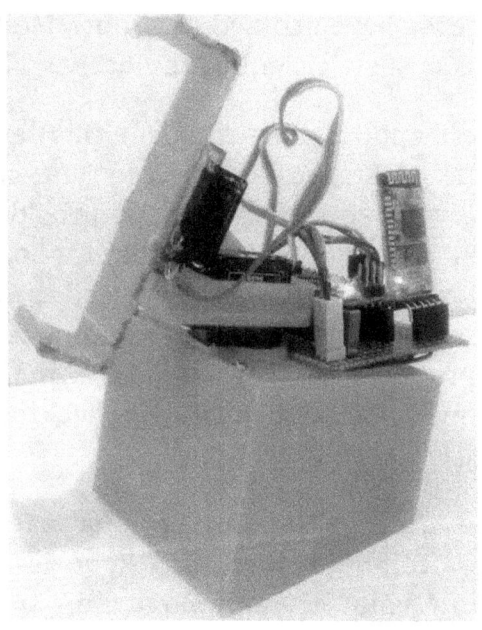

Materials Required:

- Arduino Nano

- Servo engine SG90 – 2Nos

- Android Phone with nice camera

- HC-05/HC-06 Bluetooth Module

- PC for programming

- 3D printer (discretionary)

- 9V Battery

3D Printing the Required Parts (Optional):

So as to skillet and tilt our cell phone we need some mechanical structures like a versatile holder and a couple of servo sections. You can utilize a cardboard to make one, since I have a 3D printer I chosen to 3D print these parts.

3D printing is an astounding device that can contribute a great deal when building model undertakings or trying different things with new mechanical plans. On the off chance that you have not yet found the advantages of a 3D printer or how it functions you can peruse The learners Guide to 3D printing.

In case you possess or approach a 3D printer, at that point you can utilize the STL records which can be downloaded from here to straightforwardly print and gather them. Anyway few sections like the cell phone holder may require a few alterations dependent on the elements of your telephone. I have structured it for my MOTO G cell phone. I have utilized an essential printer of mine to print every one of the parts. The printer is FABX v1 from 3ding which comes at a reasonable cost with a print volume of 10 cubic cm. The modest value accompanies an exchange off with low print goals and no SD card or print continuing capacity. I am utilizing programming called Cura to print the STL records. The settings that I used to print the materials are given underneath you can utilize the equivalent or change them dependent on your printer.

Quality	
Layer height (mm)	0.2
Shell thickness (mm)	0.8
Enable retraction	✔
Fill	
Bottom/Top thickness (mm)	0.8
Fill Density (%)	100
Speed and Temperature	
Print speed (mm/s)	50
Printing temperature (C)	190
Support	
Support type	Everywhere
Platform adhesion type	None
Filament	
Diameter (mm)	1.75
Flow (%)	90

When you print all the necessary materials you can verify them in position by utilizing screws and some heated glue. After you get together is finished it should look like beneath.

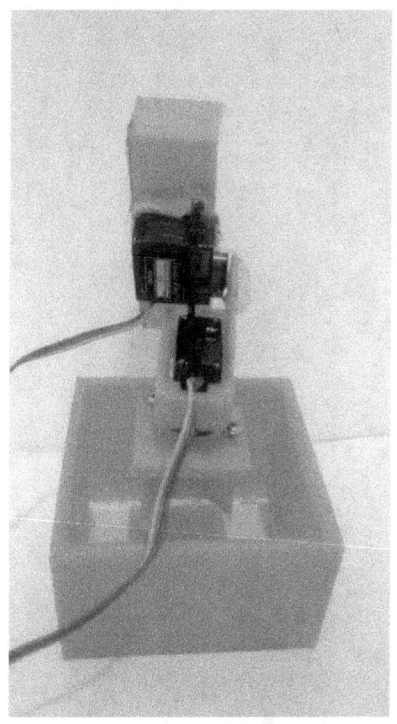

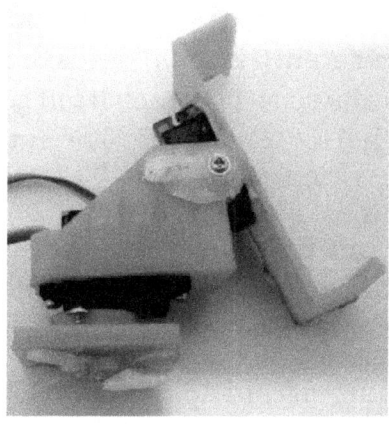

Schematic and Hardware:

The Circuit for this Face Tracking on Smart Phone task is appeared in the picture underneath:

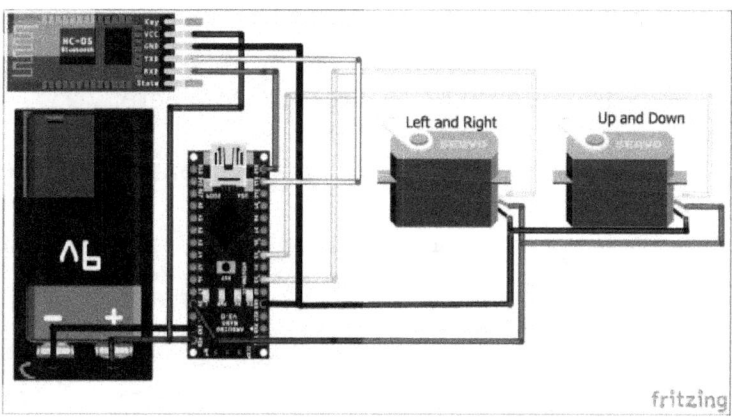

The Circuit Consists of 2 Servo engines, out of which one is utilized to move the cell phone left/right and the other is utilized to tilt the cell phone up/down. The bearing wherein the servo needs to move will be told by the Arduino Nano which itself gets data from the Bluetooth (HC-05) module. The entire circuit is controlled by a 9V battery.

This circuit can be associated effectively on your breadboard or you can likewise weld these on a little Perf board like I have done here.

Setting up your Android Application:

As, I said prior the principle mind working behind this undertaking is this Android application. This android application was created utilizing Processing Android. You can straightforwardly introduce this application on your cell phone and dispatch that by following the means underneath.

- Download the APK record from here.

- Power on the circuit appeared previously.

- In your telephone settings scan for Bluetooth module named "HC-05"

- In the event that you have named it something different other than "HC-05" transform it back to HC-05 since at exactly that point the application will work.

- Pair with your Bluetooth module with the

secret phrase "1234" or "0000".

- Presently, dispatch the Application in picture mode. You should see your camera screen and furthermore "Associated with: HC-05" on the highest point of your screen.

- Take a stab at moving your camera over a face and a green box ought to show up over it and its position will likewise be shown on the upper left corner of your screen as demonstrated as follows.

You can take this Arduino Face Tracking Project to next level by getting parcel of headways for which you won't have to code your very own Android application. Making an Android application may sound troublesome yet trust me with the assistance of Processing you can learn it in the blink of an eye. The total preparing code that is utilized to fabricate this application can be downloaded here. You are allowed to make any headway with your own innovativeness. Check beneath ventures to become familiar with Processing:

- Computer generated Reality utilizing Arduino and Processing

- Ping Pong Game utilizing Arduino

- Advanced mobile phone Controlled FM Radio utilizing Processing.

- Arduino Radar System utilizing Processing as well as Ultrasonic Sensor

Programming your Arduino:

The Android application will identify the face as well as its situation on screen; it will at that point choose which course it should move dependent on the situation of the face with the goal that the face gets to the focal point of the screen. This heading is then sent to the Arduino through Bluetooth Module.

The Arduino program for this undertaking is genuinely straightforward, we simply need to control the course of the two servo engines dependent on the qualities got from the Bluetooth Module. The total code can be found toward the finish of this instructional exercise, I have likewise clarified couple of significant lines beneath.

Beneath line of code sets up a sequential association with pins D12 as RX and D11 as TX. Thus the stick D12 must be associated with the TX of the BT module and the stick D11 to the RX of the BT module.

```
SoftwareSerial cam_BT(12, 11); // RX, TX
```

At that point we have initialised the Bluetooth module at a baud pace of 9600. Ensure you module likewise chips away at a similar baud rate. Else change it as needs be.

```
cam_BT.begin(9600); //start the Bluetooth communication at 9600 baudrate

cam_BT.println("Ready to take commands");
```

Underneath line peruses what is coming in through the Bluetooth module. Likewise the information is spared in the variable "BluetoothData".

```
if (cam_BT.available()) //Read whats coming in through Bluetooth

{
```

```
BluetoothData=cam_BT.read();

Serial.print("Incoming from BT:");

Serial.println(BluetoothData);

}
```

In view of the information got from the Bluetooth the engines course is controlled. To turn an engine left the engine is decrement by an estimation of 2 from its past position. You can build this worth 2 to 4 or 6 on the off chance that you need the arm to move quicker. Be that as it may, it may make a few bastards making the camera insecure.

```
if(BluetoothData==49) //Turn Left

{pos1+=2; servo1.write(pos1);}

if(BluetoothData==50) //Turn Right

{pos1-=2; servo1.write(pos1);}

if(BluetoothData==51) //Turn Up

{pos2-=2; servo2.write(pos2);}

if(BluetoothData==52) //Turn Down
```

```
{pos2+=2; servo2.write(pos2);}
```

Working:

When we are prepared with our equipment, code and Android Application its time for some activity. Just power your Arduino and open the android application. The Application will naturally associate with the HC-05 (must be named HC-05) Bluetooth module and will trust that a face will be identified. Basically place the telephone in our versatile holder and sit before it. You should see your servo engines moving your telephone so your face will be put at the focal point of the screen. Presently move around inside the scope of the camera and your cell phone will pursue your developments. You can likewise attempt it by setting and moving any image.

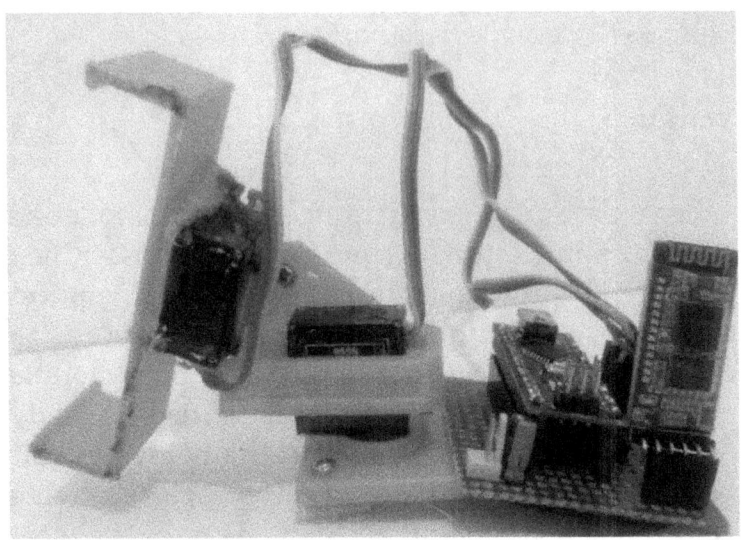

You can construct significantly more over it which is left for your imagination, I trust you delighted in the venture and made it work.

Code

/*Arduino Code for Face Tracking Arduino

*/
/*CONNECTION DETIALS
* Arduino D11 -> RX of BT Module
* Arduino D12 -> Tx of BT
* Servo1 -> pin 3 of arduino Nano to pan
* Servo2 -> pin 5 of arduino Nano to tilt
*/
#include <Servo.h> //header to drive servo motors
#include <SoftwareSerial.h>// import the serial li-

```
brary
SoftwareSerial cam_BT(12, 11); // RX, TX
int ledpin=13; //led on D13 will show blink on / off
int BluetoothData; // the data given from Computer
//lets declare the servo objects
Servo servo1;
Servo servo2;
long gmotor,gnum1,gnum2;
int pos;
int flag=0;
int pos1 = 40;
int pos2 = 90;
void setup() {
  servo1.attach(3);
  servo2.attach(5);;
  //**Initial position of all four servo motors**//
  servo1.write(pos1);
  servo2.write(pos2);
  //**initialised**//


cam_BT.begin(9600); //start the Bluetooth commu-
nication at 9600 baudrate
cam_BT.println("Ready to take commands");
Serial.begin(57600);
Serial.println("Face tracking");
}
//***Function for each Servo actions**//
void call(int motor, int num1, int num2) // The
values like Motor number , from angle and to angle
```

```
are received
{
Serial.println("Passing values...");
flag =0;
switch (motor)
 {
  case 1:      // For motor one
  {
  Serial.println("Executing motor one");
  if(num1<num2) // Clock wise rotation
  {
    for ( pos =num1; pos<=num2; pos+=1)
      {
      servo1.write(pos);
      delay( 20);
      }
  }


    if(num1>num2) // Anti-Clock wise rotation
  {
    for ( pos =num1; pos>=num2; pos-=1)
    {
    servo1.write(pos);
    delay( 20);
    }
  }
  break;
  }
    /////////JUST DUPLICATE FOR OTHER SERVOS////
```

```
case 2:  // For motor 2
{
Serial.println("Executing motor two");
if(num1<num2)
{
  for ( pos =num1; pos<=num2; pos+=1)
  {
  servo2.write(pos);
  delay( 20);
  }
}
if(num1>num2)
{
  for ( pos =num1; pos>=num2; pos-=1)
  {
  servo2.write(pos);
  delay( 20);
  }
}
  break;
  }
 }
}
void loop() {
  if(Serial.available()>0) //Read whats coming in
through Serial
  {
 gmotor= Serial.parseInt();
```

```
Serial.print(" selected Number-> ");
Serial.print(gmotor);
Serial.print(" , ");
gnum1 = Serial.parseInt();
Serial.print(gnum1);
Serial.print(" degree , ");
gnum2 = Serial.parseInt();
Serial.print(gnum2);
Serial.println(" degree ");
flag=1;
}


if (cam_BT.available()) //Read whats coming in through Bluetooth
{
BluetoothData=cam_BT.read();
Serial.print("Incoming from BT:");
Serial.println(BluetoothData);
}
if (flag ==1)
  call(gmotor,gnum1,gnum2); //call the respective motor for action
if(BluetoothData==49)//Turn Left
{pos1+=2; servo1.write(pos1);}
if(BluetoothData==50)//Turn Right
{pos1-=2; servo1.write(pos1);}
if(BluetoothData==51)//Turn Up
{pos2-=2; servo2.write(pos2);}
if(BluetoothData==52)//Turn Down
```

```
{pos2+=2; servo2.write(pos2);}
flag=0;
BluetoothData=0;
}
```

4. SEISMIC TREMOR DETECTOR ALARM UTILIZING ARDUINO

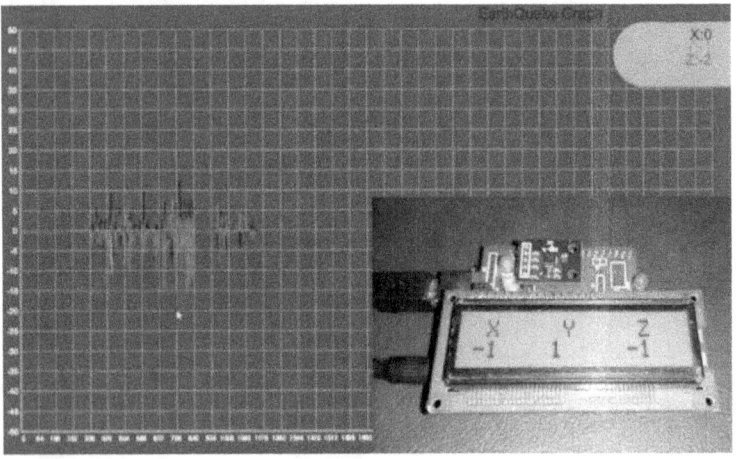

A seismic tremor is an eccentric cataclysmic event that makes harm lives and property. It happens all of a sudden and we can't stop it however we can be alarmed from it. In the present time, there are numerous innovations which can be utilized to distinguish the little shakes and thumps, with the goal that we can avoid potential risk before some significant vibrations in earth. Here we are utilizing Accelerometer ADXL335 to identify the pre-tremor vibrations. Accelerometer ADXL335 is profoundly delicate to shakes and vibrations alongside all the three tomahawks. Here we are building an Arduino based Earthquake Detector utilizing Accelerometer.

We are here building this Earthquake identifier as an Arduino Shield on PCB and will likewise show the Vibrations Graph on PC utilizing Processing.

Components Required:

- Arduino UNO
- Accelerometer ADXL335
- 16x2 LCD
- Buzzer
- BC547 transistor
- 1k Resistors
- 10K POT
- LED
- Power Supply 9v/12v
- Berg sticks male/female

Accelerometer:

Stick Description of accelerometer:

- Vcc 5 volt supply ought to interface at this stick.

- X-OUT This stick gives an Analog yield in x course

- Y-OUT This stick give an Analog Output in y course

- Z-OUT This stick gives an Analog Output in z heading

- GND Ground

- ST This stick utilized for set affectability of sensor

Likewise check our different activities utilizing Ac-

celerometer:

- Ping Pong Game utilizing Arduino

- Accelerometer Based Hand Gesture Controlled Robot.

- Arduino Based Vehicle Accident Alert System utilizing GPS, GSM as well as Accelerometer

Working Explanation:

Working of this Earthquake Detector is straightforward. As we referenced before that we have utilized Accelerometer for distinguishing seismic tremor vibrations along any of the three tomahawks so that at whatever point vibrations happen accelerometer detects that vibrations as well as convert them into comparable ADC esteem. At that point these ADC esteems are perused by Arduino and appeared over the 16x2 LCD. We have additionally indicated these qualities on Graph utilizing Processing. Study Accelerometer by experiencing our other Accelerometer extends here.

First we have to adjust the Accelerometer by taking the examples of encompassing vibrations at whatever point Arduino Powers up. At that point we have to subtract those example esteems from the real readings to get the genuine readings. This adjustment is required so it won't show alarms as for its ordinary encompassing vibrations. Subsequent to discovering genuine readings, Arduino contrasts these qualities and predefined max and min esteems. In case Arduino finds any progressions esteems are increasingly, at that point or less then the predefined estimations of any hub both way (negative and positive) at that point Arduino trigger the ringer and shows the status of caution over the 16x2 LCD and a LED turned on also. We can alter the affectability of Earthquake indicator by changing the Predefined esteems in Arduino code.

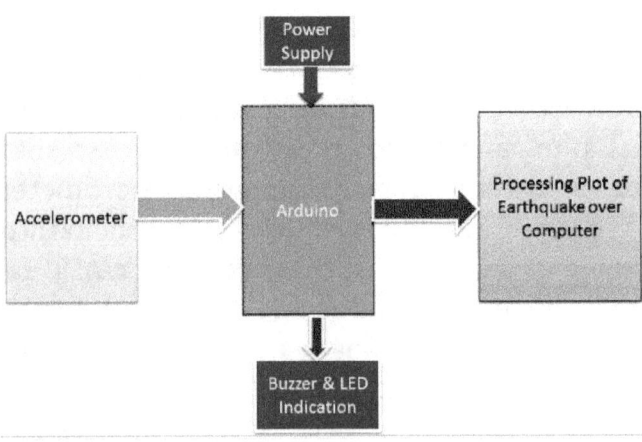

Arduino Code is given toward the finish of the article.

Circuit Explanation:

Circuit of this Earthquake finder Arduino Shield PCB is likewise straightforward. In this task, we have utilized Arduino that peruses accelerometer's simple voltage and convert them into the advanced qualities. Arduino likewise drives the bell, LED, 16x2 LCD and figure and analyze qualities and make proper move. Next part is Accelerometer which recognizes vibration of earth and creates simple voltages in 3 tomahawks (X, Y, as well as Z). LCD is utilized for indicating X, Y and Z hub's adjustment in qualities and furthermore demonstrating alarm message over it. This LCD is appended to Arduino in 4-piece mode. RS, GND, and EN pins are straightforwardly associated with 9, GND and 8 pins of Arduino and rest of 4 information pins of LCD specifically D4, D5, D6 as well as D7 are legitimately associated with computerized stick 7, 6, 5 and 4 of Arduino. The signal is associated with stick 12 of Arduino through a NPN BC547 transistor. A 10k pot is additionally utilized for controlling the splendor of the LCD.

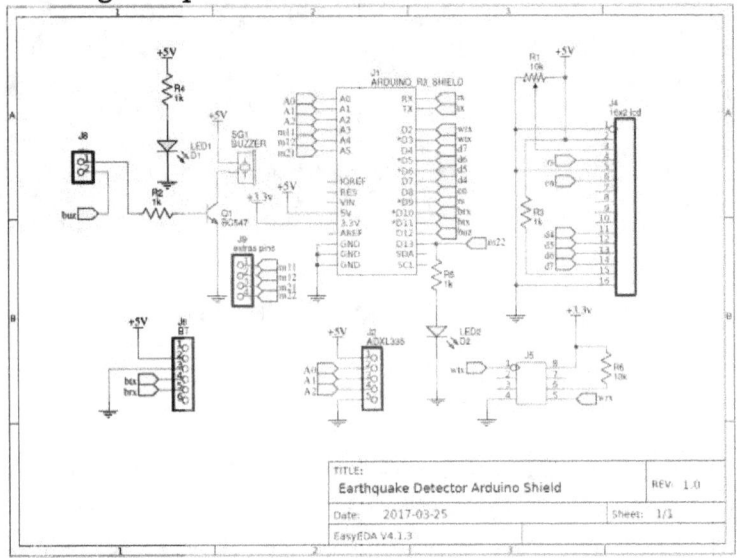

Programming Explanation:

In this Earthquake Detector Arduino Shield, we have made two codes: one for Arduino to distinguish a seismic tremor and another for Processing IDE to plot the quake vibrations over the diagram on Computer. We will find out about both the codes individually:

Arduino code:

As a matter of first importance, we adjust the accelerometer as for its putting surface, so it won't show cautions concerning its ordinary encompassing vibrations. In this adjustment, we take a few examples and afterward take a normal of them and stores in a variable.

```
for(int i=0;i<samples;i++)     // taking samples for
calibration

{

  xsample+=analogRead(x);

  ysample+=analogRead(y);

  zsample+=analogRead(z);

}

xsample/=samples;  // taking avg for x

ysample/=samples;  // taking avg for y

zsample/=samples;  // taking avg for z

delay(3000);

lcd.clear();

lcd.print("Calibrated");
```

```
delay(1000);

lcd.clear();

lcd.print("Device Ready");

delay(1000);

lcd.clear();

lcd.print("X   Y   Z ");
```

Presently at whatever point Accelerometer takes readings, we will subtract those example esteems from the readings with the goal that it can disregard surroundings vibrations.

```
int value1 = analogRead(x);  // reading x out

  int value2 = analogRead(y);  //reading y out

  int value3 = analogRead(z);  //reading z out

  int xValue = xsample-value1;  // finding change in
x

  int yValue = ysample-value2;  // finding change in
y
```

```
int zValue=zsample-value3;  // finding change in z

/*displying change in x,y and z axis values over
lcd*/

lcd.setCursor(0,1);

lcd.print(zValue);

lcd.setCursor(6,1);

lcd.print(yValue);

lcd.setCursor(12,1);

lcd.print(zValue);

delay(100)
```

At that point Arduino looks at those adjusted (subtracted) values with predefined limits. What's more, make a move in like manner. On the off chance that the qualities are higher than predefined values, at that point it will signal the bell and plot the vibration chart on PC utilizing Processing.

```
/* comparing change with predefined limits*/

if(xValue < minVal || xValue > maxVal  || yValue
```

```
< minVal || yValue > maxVal || zValue < minVal ||
zValue > maxVal)

    {

    if(buz == 0)

    start=millis();  // timer start

    buz=1;    // buzzer / led flag activated

    }

    else if(buz == 1)      // buzzer flag activated then
alerting earthquake

    {

    lcd.setCursor(0,0);

    lcd.print("Earthquake Alert  ");

      if(millis()>= start+buzTime)

    buz=0;

    }
```

Preparing code:

The following is the Processing Code joined, you can install the code from beneath connect:

Earth Quake Detector Processing Code

We have structured a diagram utilizing Processing, for earth tremor vibrations, in which we characterized the size of the window, units, text dimension, foundation, perusing and showing sequential ports, open chosen sequential port and so on.

```
// set the window size: and Font size

  f6 = createFont("Arial",6,true);

  f8 = createFont("Arial",8,true);

  f10 = createFont("Arial",10,true);

  f12 = createFont("Arial",12,true);

  f24 = createFont("Arial",24,true);

  size(1200, 700);

// List all the available serial ports

println(Serial.list());
```

```
myPort = new Serial(this, "COM43", 9600);

println(myPort);

myPort.bufferUntil('\n');

background(80)
```

In underneath work, we have gotten information from sequential port and concentrate required information and afterward mapped it with the size of the chart.

```
// extracting all required values of all three axis:

int l1 =inString.indexOf("x =")+2;

String temp1 =inString.substring(l1,l1 +3);

l1 =inString.indexOf("y =")+2;

String temp2 =inString.substring(l1,l1 +3);

l1 =inString.indexOf("z =")+2;

String temp3 =inString.substring(l1,l1 +3);

//mapping x, y and z value with graph dimensions
```

```
float inByte1 = float(temp1+(char)9);

inByte1 = map(inByte1, -80,80, 0, height-80);

float inByte2 = float(temp2+(char)9);

inByte2 = map(inByte2,-80,80, 0, height-80);

float inByte3 = float(temp3+(char)9);

inByte3 = map(inByte3,-80,80, 0, height-80);

float x=map(xPos,0,1120,40,width-40);
```

After this, we have plotted unit space, max and min limits, estimations of x, y and z-hub.

```
//ploting graph window, unit

  strokeWeight(2);

  stroke(175);

  Line(0,0,0,100);

  textFont(f24);

  fill(0,00,255);
```

```
textAlign(RIGHT);

xmargin("EarthQuake Graph",200,100);

fill(100);

strokeWeight(100);

line(1050,80,1200,80);

.... ....

..........
```

After this we plot the qualities over the diagram by utilizing 3 distinct hues as Blue for x-hub esteem, green shading for y hub and z is spoken to by red shading.

```
stroke(0,0,255);

if(y1 == 0)

y1=height-inByte1-shift;

line(x, y1, x+2, height-inByte1-shift);

y1=height-inByte1-shift;
```

```
stroke(0,255,0);

if(y2 == 0)

y2=height-inByte2-shift;

line(x, y2, x+2, height-inByte2-shift);

y2=height-inByte2-shift;

stroke(255,0,0);

if(y2 == 0)

y3=height-inByte3-shift;

line(x, y3, x+2, height-inByte3-shift);

y3=height-inByte3-shift;
```

Likewise get familiar with preparing by experiencing our other Processing ventures.

Circuit and PCB Design using EasyEDA:

EasyEDA isn't just the one stop answer for schematic catch, circuit recreation and PCB structure, they likewise offer a minimal effort PCB Prototype and Components Sourcing administration. They as of late propelled their part sourcing administration where they have a huge load of electronic segments and clients can arrange their necessary segments alongside the PCB request.

While structuring your circuits and PCBs, you can likewise make your circuit and PCB plans open with the goal that different clients can duplicate or alter them and can take profit by there, we have additionally made our entire Circuit and PCB formats open for this Earthquake Indicator Shield for Arduino UNO, check the beneath connect:

https://easyeda.com/EarthQuake_Detector-380c29e583b14de8b407d06ab0bbf70f

The following is the Snapshot of Top layer of PCB format from EasyEDA, you can see any Layer (Top, Bottom, Topsilk, bottomsilk and so on) of the PCB by choosing the layer structure the 'Layers' Window.

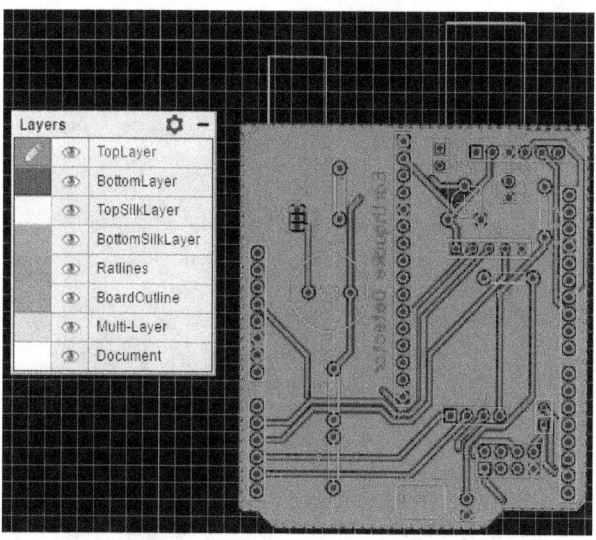

You can likewise observe Photo perspective on PCB utilizing EasyEDA:

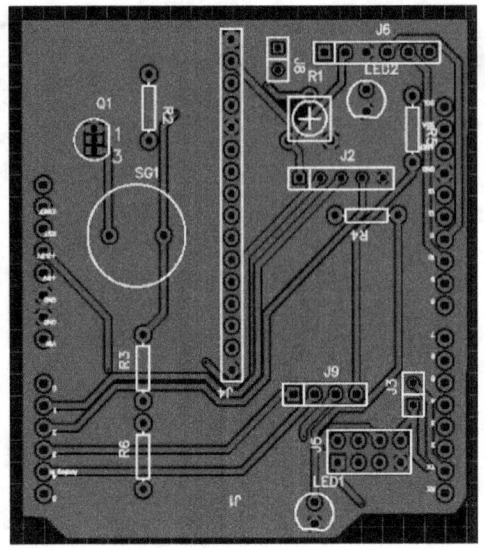

Calculating and Ordering Samples online:

In the wake of finishing the structure of PCB, you can tap the symbol of Fabrication yield, which will take you on the PCB request page. Here you can see your PCB in Gerber Viewer or install Gerber documents of your PCB. Here you can choose the quantity of PCBs you need to arrange, what number of copper layers you need, the PCB thickness, copper weight, and even the PCB shading. After you have chosen the majority of the choices, click "Spare to Cart" and complete your request. As of late they have dropped their PCB rates fundamentally and now you can arrange 10 pcs 2-layer PCB with 10cm x 10cm size only for $2.

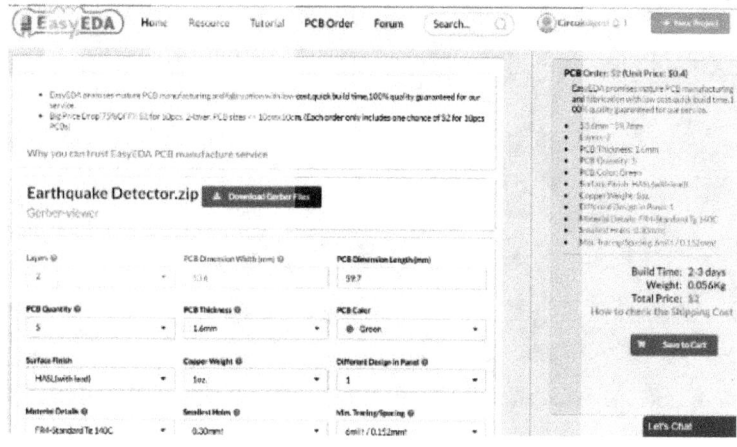

Here is the PCBs I Got from EasyEDA:

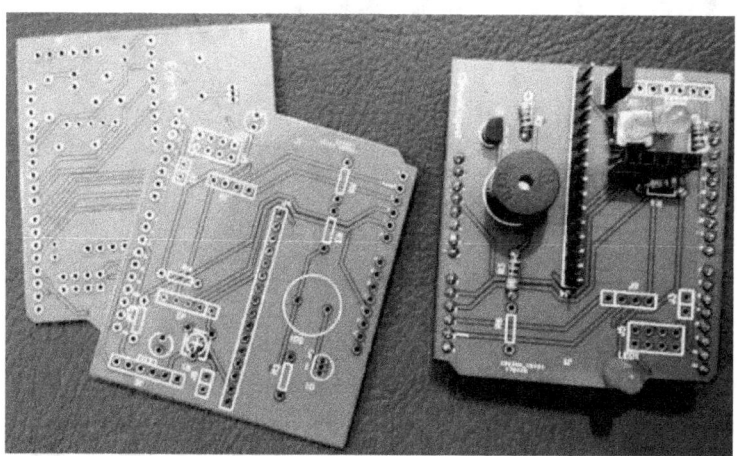

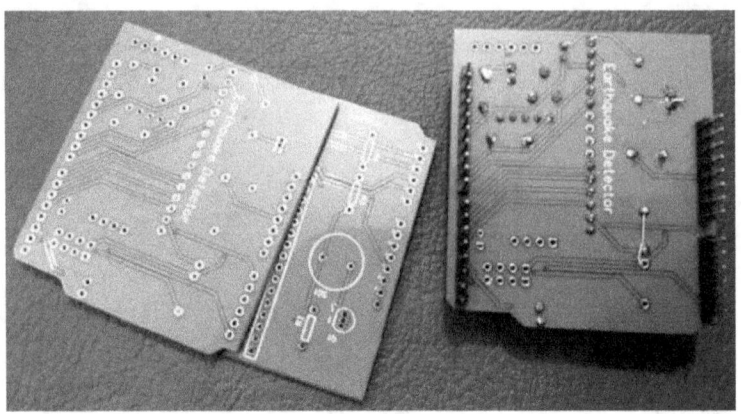

The following are the photos of definite Shield in the wake of fastening the parts on PCB:

Code

```
#include<LiquidCrystal.h>   //lcd Header
LiquidCrystal lcd(9,8,7,6,5,4);   // pins for LCD Connection
#define buzzer 12 //buzzer pin
#define led 13 //led pin
#define x A0 // x_out pin of Accelerometer
#define y A1 // y_out pin of Accelerometer
#define z A2 //z_out pin of Accelerometer
/*variables*/
int xsample=0;
int ysample=0;
int zsample=0;
long start;
int buz=0;
/*Macros*/
#define samples 50
#define maxVal 20  // max change limit
#define minVal -20   // min change limit
#define buzTime 5000 // buzzer on time
void setup()
{
 lcd.begin(16,2); //initializing lcd
 Serial.begin(9600); // initializing serial
 delay(1000);
 lcd.print("EarthQuake ");
 lcd.setCursor(0,1);
 lcd.print("Detector  ");
 delay(2000);
 lcd.clear();
 lcd.print("Hello World");
```

```
lcd.setCursor(0,1);
lcd.print("Saddam Khan  ");
delay(2000);
lcd.clear();
lcd.print("Calibrating.....");
lcd.setCursor(0,1);
lcd.print("Please wait...");
pinMode(buzzer, OUTPUT);
pinMode(led, OUTPUT);
buz=0;
digitalWrite(buzzer, buz);
digitalWrite(led, buz);
 for(int i=0;i<samples;i++)      // taking samples for
calibration
 {
 xsample+=analogRead(x);
 ysample+=analogRead(y);
 zsample+=analogRead(z);
 }
xsample/=samples;  // taking avg for x
ysample/=samples;   // taking avg for y
zsample/=samples;  // taking avg for z


 delay(3000);
lcd.clear();
lcd.print("Calibrated");
delay(1000);
lcd.clear();
lcd.print("Device Ready");
 delay(1000);
```

```
lcd.clear();
lcd.print("X  Y  Z ");
}
void loop()
{
  int value1=analogRead(x);  // reading x out
  int value2=analogRead(y);  //reading y out
  int value3=analogRead(z);  //reading z out
  int xValue=xsample-value1;  // finding change in x
  int yValue=ysample-value2;  // finding change in y
  int zValue=zsample-value3;  // finding change in z
  /*displying change in x,y and z axis values over lcd*/
  lcd.setCursor(0,1);
  lcd.print(zValue);
  lcd.setCursor(6,1);
  lcd.print(yValue);
  lcd.setCursor(12,1);
  lcd.print(zValue);
  delay(100);
  /* comparing change with predefined limits*/
  if(xValue < minVal || xValue > maxVal || yValue <
minVal || yValue > maxVal || zValue < minVal || zValue
> maxVal)
  {
   if(buz == 0)
   start=millis(); // timer start
   buz=1;   // buzzer / led flag activated
  }
  else if(buz == 1)       // buzzer flag activated then
alerting earthquake
```

```
{
  lcd.setCursor(0,0);
  lcd.print("Earthquake Alert  ");
  if(millis()>= start+buzTime)
  buz=0;
}


  else
{
  lcd.clear();
  lcd.print("X   Y   Z ");
}
  digitalWrite(buzzer, buz);    // buzzer on and off
command
  digitalWrite(led, buz); // led on and off command
  /*sending values to processing for plot over the
graph*/
  Serial.print("x=");
  Serial.println(xValue);
  Serial.print("y=");
  Serial.println(yValue);
  Serial.print("z=");
  Serial.println(zValue);
  Serial.println(" $");
}
```

5. ARDUINO TIMER TUTORIAL

The Arduino Development Platform was initially created in 2005 as a simple to-utilize programmable gadget for craftsmanship configuration ventures. Its expectation was to help non-designers to work with fundamental hardware and microcontrollers absent much by way of programming information. Be that as it may, at that point, due to its simple to utilize nature it was before long adjusted by gadgets learners and specialists around the globe and today it is even favored for model improvement and POC advancements.

While it is alright in the first place Arduino, it is imperative to gradually move into the center microcontrollers like AVR, ARM, PIC, STM and so forth and program it utilizing their local applications. This is on the grounds that the Arduino Programming language is exceptionally straightforward as the larger part of the work is finished by pre-constructed capacities like digitalWrite(), AnalogWrite(), Delay() and so forth while the low level machine language is taken cover behind them. The Arduino projects are not like other Embedded C coding where we manage register bits as well as make them high or else low dependent on the rationale of our program.

Arduino Timers immediately:

Thus, to comprehend what's going on inside the prefabricated capacities we have to burrow behind these terms. For instance when a postponement() work is utilized it genuine sets the Timer and Counter Register bits of the ATmega microcontroller.

In this arduino clock instructional exercise we will maintain a strategic distance from the utilization of this postponement() work and rather really manage the Registers themselves. The beneficial is you can utilize the equivalent Arduino IDE for this. We will set our Timer register bits and utilize the Timer

Overflow Interrupt to flip a LED each time the interfere with happens. The preloader estimation of the Timer bit can likewise be balanced utilizing pushbuttons to control the length where the intrude on happens.

What is TIMER in Embedded Electronics?

Clock is somewhat interfere. It resembles a straightforward clock which can quantify time interim of an occasion. Each microcontroller has a clock (oscillator), state in Arduino Uno it is 16Mhz. This is liable for speed. Higher the clock recurrence higher will be the preparing speed. A clock uses counter which checks at certain speed contingent on the clock recurrence. In Arduino Uno it takes 1/16000000 seconds or else 62nano seconds to make the most of a solitary. Which means Arduino moves starting with one guidance then onto the next guidance for each 62 nano second.

Clocks in Arduino UNO:

In Arduino UNO there are three clocks utilized for various capacities.

Timer0:

It is a 8-Bit clock and utilized in clock capacity, for example, delay(), millis().

Timer1:

It is a 16-Bit clock and utilized in servo library.

Timer2:

It is a 8-Bit Timer and utilized in tone() work.

Arduino Timer Registers

To change the arrangement of the clocks, clock registers are utilized.

1. Clock/Counter Control Registers (TCCRnA/B):

This register holds the principle control bits of the clock as well as used to control the prescalers of clock. It additionally permits to control the method of clock utilizing the WGM bits.

Casing Format:

TCCR1A	7	6	5	4	3	2	1	0
	COM1A1	COM1A0	COM1B1	COM1B0	COM1C1	COM1C0	WGM11	WGM10

Prescaler:

The CS12, CS11, CS10 bits in TCCR1B sets the prescaler esteem. A prescaler is utilized to arrangement the clock speed of the clock. Arduino Uno has prescalers of 1, 8, 64, 256, 1024.

CS12	CS11	CS10	USE

0	0	0	No Clock Timer STOP
0	0	1	CLCK i/o /1 No Prescaling
0	1	0	CLK i/o /8 (From Prescaler)
0	1	1	CLK i/o /64 (From Prescaler)
1	0	0	CLK i/o /256 (From Prescaler)
1	0	1	CLK i/o /1024 (From Prescaler)
1	1	0	External clock source on T1 Pin. Clock on falling edge
1	1	1	External Clock source on T1 pin. Clock on rising edge.

2. Clock/Counter Register (TCNTn)

This Register is utilized to control the counter worth and to set a preloader esteem.

Recipe for preloader esteem for required time in second:

TCNTn = 65535 – (16x1010xTime in sec/Prescaler Value)

To ascertain preloader esteem for timer1 for time of 2 Sec:

$$TCNT1 = 65535 - (16 \times 1010 \times 2/1024) = 34285$$

Arduino Timer Interrupts

We recently found out about Arduino Interrupts and have seen that Timer hinders are somewhat programming interferes. There are different clock hinders in Arduino which are clarified underneath.

Clock Overflow Interrupt:

At whatever point the clock ranges to its most extreme worth state for instance (16 Bit-65535) the Timer Overflow Interrupt happens. Along these lines, an ISR intrude on administration routine is considered when the Timer Overflow Interrupt bit empowered in the TOIEx present in clock interfere with cover register TIMSKx.

ISR Format:

ISR(TIMERx_OVF_vect)

{

}

Yield Compare Register (OCRnA/B):

Here when the Output Compare Match Interrupt happens then the interfere with administration ISR (TIMERx_COMPy_vect) is called and furthermore OCFxy banner piece will be set in TIFRx register. This ISR is empowered by setting empower bit in OCIExy present in TIMSKx register. Where TIMSKx is Timer Interrupt Mask Register.

Clock Input Capture:

Next when the clock Input Capture Interrupt happens then the interfere with administration ISR (TIMERx_CAPT_vect) is called and furthermore the ICFx banner piece will be set in TIFRx (Timer Interrupt Flag Register). This ISR is empowered by setting the empower bit in ICIEx present in TIMSKx register.

Components Required

- Arduino UNO
- Push Buttons (2)
- LED (Any Color)
- 10k Resistor (2), 2.2k (1)
- 16x2 LCD Display

Circuit Diagram

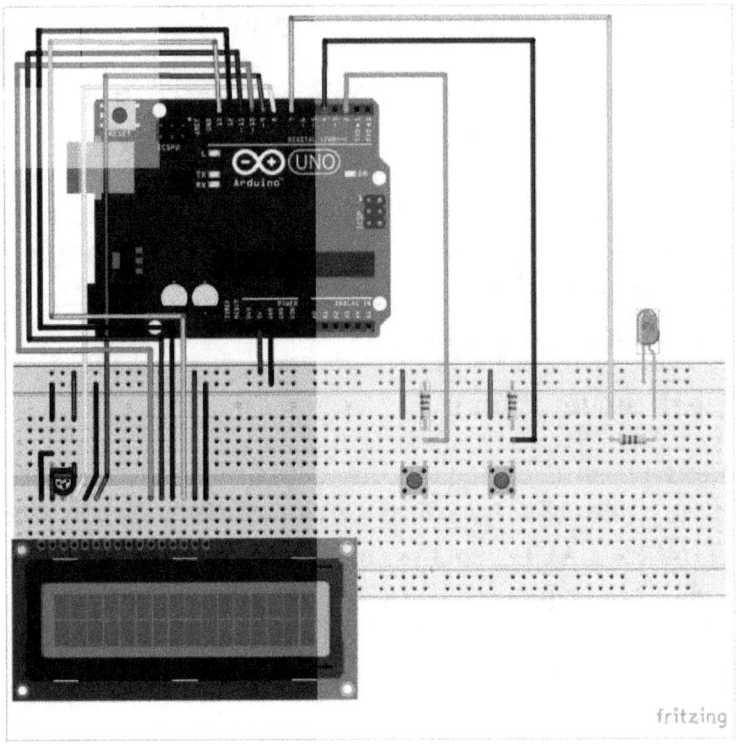

Circuit Connections between Arduino UNO and 16x2 LCD show:

16x2 LCD	Arduino UNO
VSS	GND
VDD	+5V
V0	To potentiometer centre pin for contrast control of LCD

RS	8
RW	GND
E	9
D4	10
D5	11
D6	12
D7	13
A	+5V
K	GND

Two Push catches with draw down resistors of 10K are associated with the Arduino pins 2 and 4 and a LED is associated with PIN 7 of Arduino through a 2.2K resistor.

The arrangement will look like beneath picture.

Programming Arduino UNO Timers

In this instructional exercise we will utilize the TIMER OVERFLOW INTERRUPT and use it to squint the LED ON and OFF for certain term by modifying the preloader esteem (TCNT1) utilizing pushbuttons. Complete code for Arduino Timer is given toward the end. Here we are clarifying the code line by line:

As 16x2 LCD is utilized in the task to show the preloader esteem, so fluid precious stone library is utilized.

```
#include<LiquidCrystal.h>
```

The LED anode stick that is associated with Arduino stick 7 is characterized as ledPin.

```
#define ledPin 7
```

Next the item for getting to Liquid Crystal class is pronounced with the LCD pins (RS, E, D4, D5, D6, D7) that are associated with Arduino UNO.

```
LiquidCrystal lcd(8,9,10,11,12,13);
```

At that point set the preloader esteem 3035 for 4 seconds. Check the equation above to compute the preloader esteem.

```
float value = 3035;
```

Next in void arrangement(), first set the LCD in 16x2 mode and show an invite message for few moments.

```
lcd.begin(16,2);

 lcd.setCursor(0,0);

 lcd.print("ARDUINO TIMERS");
```

```
delay(2000);

lcd.clear();
```

Next set the LED stick as OUTPUT stick and the Push catches are set as INPUT pins

```
pinMode(ledPin, OUTPUT);

pinMode(2,INPUT);

pinMode(4,INPUT);
```

Next debilitate every one of the interferes:

```
noInterrupts();
```

Next the Timer1 is instated.

```
TCCR1A = 0;

TCCR1B = 0;
```

The preloader clock worth is set (Initially as 3035).

```
TCNT1 = value;
```

At that point the Pre scaler esteem 1024 is set in the TCCR1B register.

```
TCCR1B |= (1 << CS10)|(1 << CS12);
```

The Timer flood hinder is empowered in the Timer Interrupt Mask register with the goal that the ISR can be utilized.

```
TIMSK1 |= (1 << TOIE1);
```

Finally all hinders are empowered.

```
interrupts();
```

Presently compose the ISR for Timer Overflow Interrupt which is liable for turning LED ON and OFF utilizing digitalWrite. The state changes at whatever point the clock flood hinder happens.

```
ISR(TIMER1_OVF_vect)
```

```
{

  TCNT1 = value;

  digitalWrite(ledPin, digitalRead(ledPin) ^ 1);

}
```

In the void circle() the estimation of preloader is increased or decremented by utilizing the push catch inputs and furthermore the worth is shown on 16x2 LCD.

```
if(digitalRead(2) == HIGH)

{

  value = value+10;      //Incement preload value

}

if(digitalRead(4)== HIGH)

{

  value = value-10;      //Decrement preload value

}
```

```
    lcd.setCursor(0,0);

    lcd.print(value);

}
```

So this is the means by which a clock can be utilized to deliver delay in Arduino program.

Code

```
#include<LiquidCrystal.h>      //LCD display library
#define ledPin 7
LiquidCrystal lcd(8,9,10,11,12,13);
float value = 3035;          //Preload timer value (3035
for 4 seconds)
void setup()
{
 lcd.begin(16,2);
 lcd.setCursor(0,0);
 lcd.print("ARDUINO TIMERS");
 delay(2000);
 lcd.clear();


  pinMode(ledPin, OUTPUT);
 pinMode(2,INPUT);
 pinMode(4,INPUT);


  noInterrupts();            // disable all interrupts
```

```
  TCCR1A = 0;
  TCCR1B = 0;
  TCNT1 = value;            // preload timer
  TCCR1B |= (1 << CS10)|(1 << CS12);  // 1024 prescaler
  TIMSK1 |= (1 << TOIE1);        // enable timer overflow
interrupt ISR
  interrupts();            // enable all interrupts
}
ISR(TIMER1_OVF_vect)              // interrupt service
routine for overflow
{
  TCNT1 = value;            // preload timer
    digitalWrite(ledPin, digitalRead(ledPin) ^ 1);   //
Turns LED ON and OFF
}
void loop()
{
  if(digitalRead(2) == HIGH)
  {
    value = value+10;      //Incement preload value
  }
  if(digitalRead(4)== HIGH)
  {
    value = value-10;      //Decrement preload value
  }
  lcd.setCursor(0,0);
  lcd.print(value);
}
```

6. ARDUINO BASED BLUETOOTH BIPED BOB (WALKING AND DANCING ROBOT)

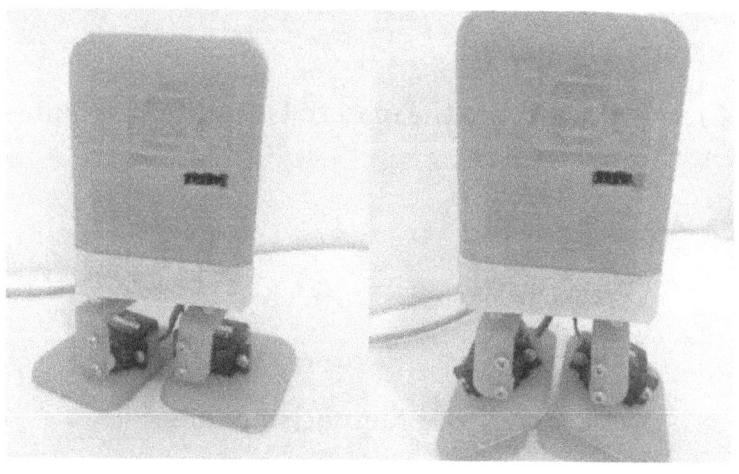

Welcome to another undertaking in which we will construct a little Robot which can walk and move. The undertaking points in showing you how to make little interest robots utilizing Arduino and how to program your Servo engines for such applications. Toward the finish of the venture you will have the option to make this strolling and moving robot that takes order from an Android Mobile Phone to play out some pre-characterized activities. You can like-wise utilize the program (given toward the finish of the instructional exercise) to effectively control the activities of your own special robot by controlling the situation of the servo engines utilizing the Ser-ial screen. Having a 3d printer will make this task all the more fascinating and look cool. In case, you don't have one you can use any of the online administra-tions or simply utilize some cardboard to construct the equivalent.

Materials Required:

Coming up next are the materials required for build-ing this robot:

- Arduino nano
- Male berg sticks
- Servo SG90 - 4Nos
- 3D printer
- HC-05/HC-06 Bluetooth module

As should be obvious this 3D printed robot requires exceptionally negligible gadgets parts to work to keep the expense of the venture as low as could be expected under the circumstances. This undertaking is just for theoretical and fun reason and doesn't have any constant application up until this point.

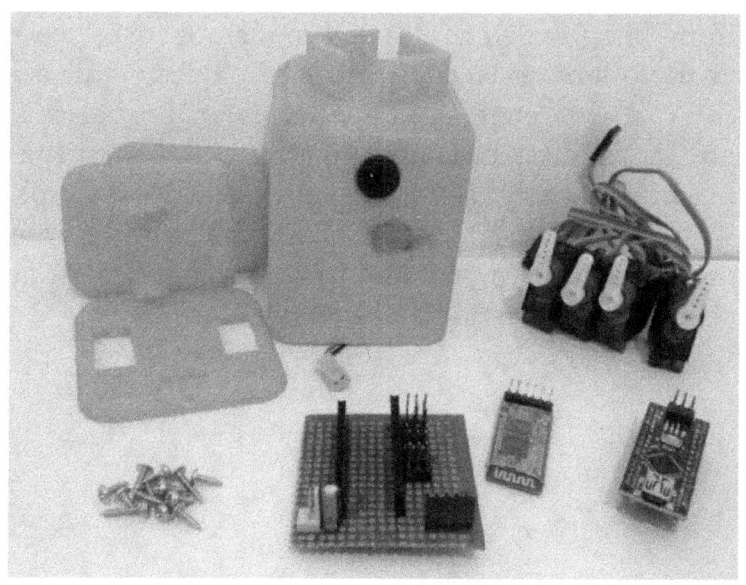

3D printing the required parts:

3D printing is an astonishing device that can contribute a great deal when building model ventures or exploring different avenues regarding new mechanical structures. In the event that you have not yet found the advantages of a 3D printer or how it functions you can peruse The amateurs Guide to 3D printing.

In this task the body of the Robot appeared above is totally 3D printed. You can download the STL documents from here. Burden these documents on your 3D printing programming like Cura and legitimately print them. I have utilized an essential printer of mine to print every parts. The printer is FABX v1 from 3ding which comes at a moderate cost with a print volume of 10 cubic cm. The modest value accompanies an exchange off with low print goals and no SD card or print continuing capacity. I am utilizing programming called Cura to print the STL records. The settings that I used to print the materials are given underneath you can utilize the equivalent or change them dependent on your printer.

Quality

Layer height (mm)	0.2
Shell thickness (mm)	0.8
Enable retraction	✔

Fill

Bottom/Top thickness (mm)	0.8
Fill Density (%)	25

Speed and Temperature

Print speed (mm/s)	35
Printing temperature (C)	190

Support

Support type	Touching buildplate
Platform adhesion type	None

Filament

Diameter (mm)	1.75
Flow (%)	90

When you print every parts clean the backings (assuming any) and afterward ensure the openings on the leg and stomach part are enormous enough to fit a screw. If not, utilize a needle to make the gap softly greater. Your 3D printed parts will look like something beneath.

Hardware and Schematics:

The Hardware for this Mobile Phone Controlled Biped Arduino Robot is extremely basic. The total schematics is appeared in the underneath picture

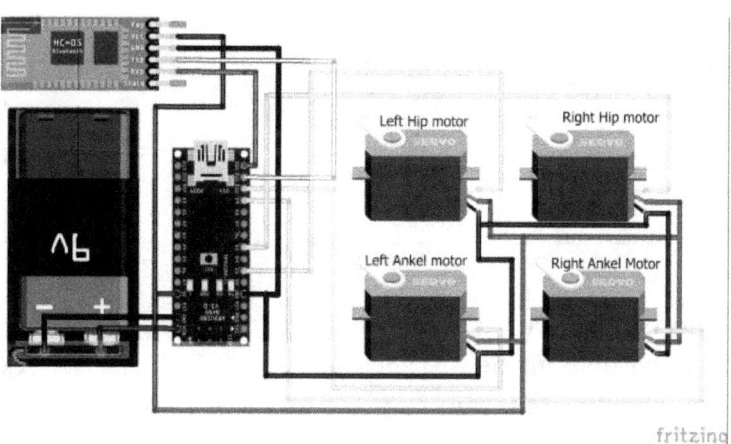

I have utilized a Perf board to make the above associations. Ensure that your circuit will likewise fit inside the leader of the robot. When your Perf board is prepared it should look something like underneath.

Assembling the robot:

When the Hardware as well as the 3D printed parts are prepared we can amass the robot. Before fixing the engines ensure you place the engines in the underneath points with the goal that the program works perfectly.

Motor Number	Motor place	Motor position
1	Left Hip motor	110
2	Right Hip motor	100
4	Right Ankle Motor	90
5	Right Hip motor	80

These edges can be set by utilizing the program given at the finish of the instructional exercise. Essentially transfer the program to your Arduino in the wake of making the above associations and type in the accompanying in the sequential screen (Note: Baud rate is 57600).

1, 100, 110

2,90,100

4,80,90

5,70,80

Your Serial screen should look something like this subsequent to putting every one of your engines in position.

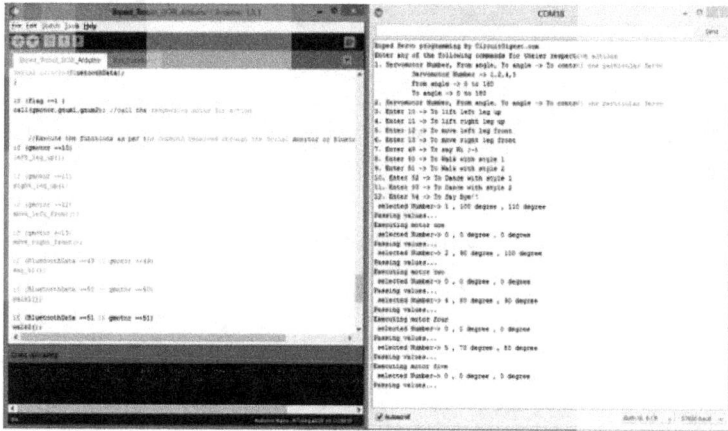

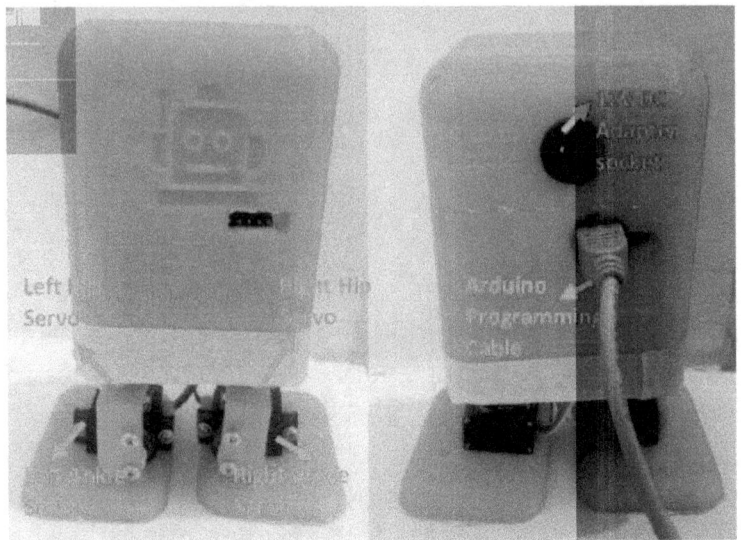

When the engines are set in the relating edges mount them as appeared in above figure.

When the Robot is amassed the time has come to program our moving robot

Programming the Arduino for Biped Robot:

Programming the BBB Robot is the most intriguing and fun part in this instructional exercise. In case you are excellent in programming servo engines with Arduino, at that point I would prescribe you to make your program. Bt, in the event that you need figure out how to utilize servo engines for mechanical applications like this, at that point this program will be exceptionally useful for. You can become familiar with arduino programming in our arduino ventures classification.

The total program is given toward the finish of this instructional exercise, or you can download the total code from here. I will clarify the sections of the equivalent underneath. The program is equipped for controlling the Robots activities through sequential screen or Bluetooth. You can likewise make your own moves by controlling each individual engine utilizing the sequential screen.

```
servo1.attach(3);

servo2.attach(5);

servo4.attach(9);

servo5.attach(10);
```

The above lines of code it use to make reference to which servo engine is associated with which stick of the Arduino. Here for our situation Servo 1,2,4 and 5 are associated with pins 3,5,9 and 10 separately.

```
Bot_BT.begin(9600); //start the Bluetooth communication at 9600 baudrate
```

```
Serial.begin(57600);
```

As said before our strolling robot can chip away at Bluetooth directions and furthermore from directions from the sequential screen. Subsequently the Bluetooth sequential correspondence works with a Baud Rate of 9600 and the sequential correspondence works with Baud Rate of 57600. The name of our Bluetooth object here is "Bot_BT".

```
switch (motor)

{

  case 1:      // For motor one

    { Serial.println("Executing motor one");

      if(num1<num2) // Clock wise rotation

  { for ( pos =num1; pos<=num2; pos+=1)

  {

    servo1.write(pos);

    delay( 20);

  }}
```

```
if(num1>num2) // Anti-Clock wise rotation

{

for ( pos =num1; pos>=num2; pos-=1)

{

  servo1.write(pos);

  delay( 20);

}}

   break;

  }

    ////////JUST        DUPLICATE    FOR    OTHER
SERVOS////

  case 2:  // For motor 2

  {

    Serial.println("Executing motor two");

    if(num1<num2)

{
```

```
for ( pos =num1; pos<=num2; pos+=1)

{

  servo2.write(pos);

  delay( 20);

}}

if(num1>num2)

{

for ( pos =num1; pos>=num2; pos-=1)

{

  servo2.write(pos);

  delay( 20);

}}

   break;

   }

  case 4:  // for motor four
```

```
    {

      Serial.println("Executing motor four");

      if(num1<num2)

  {

  for ( pos =num1; pos<=num2; pos+=1)

  {

    servo4.write(pos);

    delay (20);

  }}

if(num1>num2)

  {

  for ( pos =num1; pos>=num2; pos-=1)

  {

    servo4.write(pos);

    delay (20);
```

```
}}
    break;

    }

    case 5:  // for motor five

    {

      Serial.println("Executing motor five");

{

for ( pos =num1; pos<=num2; pos+=1)

    if(num1<num2)

{

  servo5.write(pos);

  delay (20);

}}

if(num1>num2)

  {
```

```
for ( pos =num1; pos> =num2; pos-=1)

{

  servo5.write(pos);

  delay (20);

}}

    break;

  }
```

The switch case appeared above is utilized to control the servo engines independently. This will help in making your own innovative moves with your robot. With this section of code you can just tell the engine number, from edge and to edge to make a specific engine move to an ideal area.

For instance on the off chance that we need to move the engine number 1 which is the left hip engine from its default area of 110 degree to 60 degree. We can basically state "1,110,60" in the sequential screen of Arduino and hit enter. This will prove to be useful to make your own mind boggling moves with your Robot. When you explore different avenues regarding all the from heavenly attendant and to point you would then be able to make your own moves and rehash them by making it as a capacity.

```
if(Serial.available()>0) //Read whats coming in
through Serial

  {

  gmotor= Serial.parseInt();

  Serial.print(" selected Number-> ");

  Serial.print(gmotor);

  Serial.print(" , ");

  gnum1 = Serial.parseInt();

  Serial.print(gnum1);

  Serial.print(" degree , ");

  gnum2 = Serial.parseInt();

  Serial.print(gnum2);

  Serial.println(" degree ");

  flag=1;

  }
```

In case a Serial information is accessible the number before the first "," is considered as gmotor and afterward the number before the second "," is considered as gnum1 as well as the number after the second "," is considered as gnum2.

```
if (Bot_BT.available()) //Read whats coming in through Bluetooth

{

BluetoothData=Bot_BT.read();

Serial.print("Incoming from BT:");

Serial.println(BluetoothData);

}
```

On the off chance that the Bluetooth gets some data, the got data is put away in the variable "Bluetooth-Data". This variable is then contrasted with the pre-characterized qualities to execute a specific activity.

```
if(flag ==1 )

call(gmotor,gnum1,gnum2); //call the respective motor for action
```

```
//Execute the functions as per the commond re-
ceived through the Serial monitor or Bluetooth//

if(gmotor ==10)

left_leg_up();

if(gmotor ==11)

right_leg_up();

if(gmotor ==12)

move_left_front();

if(gmotor ==13)

move_right_front();

if(BluetoothData ==49 || gmotor ==49)

say_hi();

if(BluetoothData ==50 || gmotor ==50)

walk1();

if(BluetoothData ==51 || gmotor ==51)

walk2();
```

```
if(BluetoothData ==52 || gmotor ==52)

dance1();

if(BluetoothData ==53 || gmotor ==53)

dance2();

if(BluetoothData ==54 || gmotor ==54)

{test();test();test();}
```

This is the place the capacities are called dependent on the qualities got from the sequential screen or the Bluetooth. As appeared over the variable gmotor will have the estimation of sequential screen and BluetoothData will have the incentive from Bluetooth gadget. The numbers 10,11,12 upto 53,54 are precharacterized numbers.

For instance on the off chance that you enter the number 49 in the sequential screen. The say_hi() capacity will be executed where the robot will wave you a howdy.

Every one of the capacities are characterized inside the page "Bot_Functions". You can open it and see what really occurs inside each capacity. Every one of these capacities were made by testing th e from blessed messenger and to heavenly attendant of each engine utilizing the switch case clarified previously.

Processing based Android Application:

The Android application to control the Robot was assemble utilizing the Processing Android mode. On the off chance that you need to roll out certain improvements to the Application you can download the total Processing program from here.

In case you basically need to utilize the application you can download it from here as an APK record and straightforwardly introduce it on your cell phone.

Note: Your Bluetooth module ought to be named HC-06 else the application won't have the option to associate with your Bluetooth Module.

When the application is introduced, you can match thea Bluetooth module with your Phone and afterward dispatch the application. It should look like underneath.

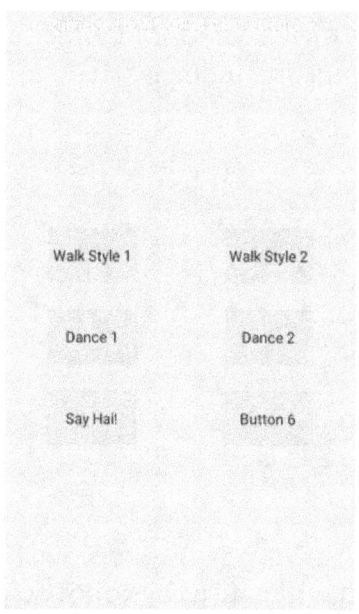

In case you need to make your application increasingly appealing or associate with some other gadget other than Hc-06. You can utilize the preparing code and roll out certain improvements to it and afterward transfer the code legitimately to your telephone.

Working of Bluetooth Controlled Biped Robot:

When your Hardware, Android Application and Arduino Sketch is prepared the time has come to have a great time with our robot. You can control the Robot from Bluetooth Application by utilizing the catches in the application or legitimately from Serial screen by utilizing as of the accompanying directions as appeared in the picture underneath.

```
Enter any of the following commands for theier respective actions
1. Servomotor Number, From angle, To angle -> To control one particular Servo
          Servomotor Number -> 1,2,4,5
          From angle -> 0 to 180
          To angle -> 0 to 180
2. Servomotor Number, From angle, To angle -> To control one particular Servo
3. Enter 10 -> To lift left leg up
4. Enter 11 -> To lift right leg up
5. Enter 12 -> To move left leg front
6. Enter 13 -> To move right leg front
7. Enter 49 -> To say Hi ;-)
8. Enter 50 -> To Walk with style 1
9. Enter 51 -> To Walk with style 2
10. Enter 52 -> To Dance with style 1
11. Enter 53 -> To Dance with style 2
12. Enter 54 -> To Say Bye!!
```

Each direction will cause the robot to play out some unconventional assignments and you can likewise add on more activities dependent on your innovativeness.

The Robot can likewise be fueled by a 12V connector or can likewise be controlled by utilizing a 9V battery. This battery can be effectively situated underneath the Perfboard and can likewise be secured with the Head of the Robot.

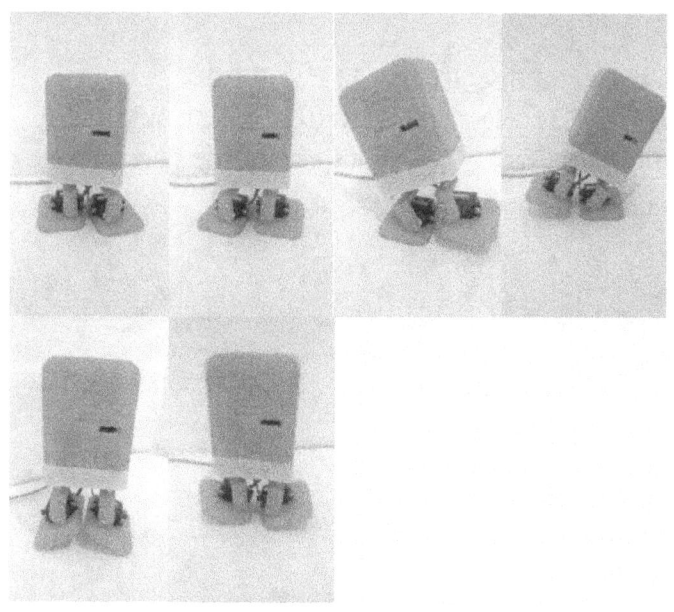

Code

/*Arduino Code for Walking and Dancing Robot
*/

/*CONNECTION DETIALS
 * Arduino D11 -> RX of BT Module
 * Arduino D12 -> Tx of BT
 * Arduino D2 -> Hall sensor 3rd pin
 * Servo1 -> pin 3 of arduino Nano
 * Servo2 -> pin 5 of arduino Nano
 * Servo4 -> pin 9 of arduino Nano
 * Servo5 -> pin 10 of arduino Nano
 */

#include <Servo.h> //header to srive servo motors
#include <SoftwareSerial.h>// import the serial li-

```
brary
SoftwareSerial Bot_BT(12, 11); // RX, TX
int ledpin=13; // led on D13 will show blink on / off
int BluetoothData; // the data given from Computer
//lets declare the servo objects
Servo servo1;
Servo servo2;
Servo servo3;
Servo servo4;
Servo servo5;
//End of declaration
long gmotor,gnum1,gnum2;
int pos,pos2;
int flag=0;
int poss1,poss2,poss3,poss4;
void setup()
{
  servo1.attach(3);
  servo2.attach(5);;
  servo4.attach(9);
  servo5.attach(10);
  //**Initial position of all four servo motors**//
  servo1.write(110);
  servo2.write(100);
  servo4.write(90);
  servo5.write(80);
  //**inititialised**//


Bot_BT.begin(9600); //start the Bluetooth communi-
```

cation at 9600 baudrate

```
Bot_BT.println("Blue Bob is ready to take actions");
Serial.begin(57600);
Serial.println("Biped Servo");
Serial.println("Enter any of the following commands for theier respective actions");
Serial.println("1. Servomotor Number, From angle, To angle -> To control one particular Servo");
Serial.println("      Servomotor Number -> 1,2,4,5");
Serial.println("      From angle -> 0 to 180");
Serial.println("      To angle -> 0 to 180");
Serial.println("2. Servomotor Number, From angle, To angle -> To control one particular Servo");
Serial.println("3. Enter 10 -> To lift left leg up");
Serial.println("4. Enter 11 -> To lift right leg up");
Serial.println("5. Enter 12 -> To move left leg front");
Serial.println("6. Enter 13 -> To move right leg front");
Serial.println("7. Enter 49 -> To say Hi ;-)");
Serial.println("8. Enter 50 -> To Walk with style 1");
Serial.println("9. Enter 51 -> To Walk with style 2");
Serial.println("10. Enter 52 -> To Dance with style 1");
Serial.println("11. Enter 53 -> To Dance with style 2");
Serial.println("12. Enter 54 -> To Say Bye!!");
}
//***Function for each Servo actions**//
void call(int motor, int num1, int num2) // The values like Motor number , from angle and to angle are received
{
```

```
Serial.println("Passing values...");
flag =0;
switch (motor)
 {
  case 1:      // For motor one
   { Serial.println("Executing motor one");
    if(num1<num2) // Clock wise rotation
{ for ( pos =num1; pos<=num2; pos+=1)
{
 servo1.write(pos);
 delay( 20);
}}
if(num1>num2) // Anti-Clock wise rotation
 {
 for ( pos =num1; pos>=num2; pos-=1)
 {
  servo1.write(pos);
  delay( 20);
}}
   break;
   }
   ////////JUST DUPLICATE FOR OTHER SERVOS////


  case 2:  // For motor 2
   {
    Serial.println("Executing motor two");
    if(num1<num2)
 {
 for ( pos =num1; pos<=num2; pos+=1)
```

```
{
 servo2.write(pos);
 delay( 20);
}}
if(num1>num2)
 {
 for ( pos =num1; pos>=num2; pos-=1)
 {
 servo2.write(pos);
 delay( 20);
}}
   break;
   }



   case 4:  // for motor four
   {
   Serial.println("Executing motor four");
   if(num1<num2)
{
 for ( pos =num1; pos<=num2; pos+=1)
 {
 servo4.write(pos);
 delay (20);
}}
if(num1>num2)
 {
 for ( pos =num1; pos>=num2; pos-=1)
 {
```

```
  servo4.write(pos);
  delay (20);
}}
  break;
  }


    case 5:  // for motor five
  {
    Serial.println("Executing motor five");
{
for ( pos =num1; pos<=num2; pos+=1)
   if(num1<num2)
{
 servo5.write(pos);
  delay (20);
}}
if(num1>num2)
 {
 for ( pos =num1; pos>=num2; pos-=1)
 {
  servo5.write(pos);
  delay (20);
}}
  break;
  }
 }
}
void loop()
{
```

```
if(Serial.available()>0) //Read whats coming in through Serial
{
gmotor= Serial.parseInt();
Serial.print(" selected Number-> ");
Serial.print(gmotor);
Serial.print(" , ");
gnum1 = Serial.parseInt();
Serial.print(gnum1);
Serial.print(" degree , ");
gnum2= Serial.parseInt();
Serial.print(gnum2);
Serial.println(" degree ");
flag=1;
}
 if (Bot_BT.available()) //Read whats coming in through Bluetooth
{
BluetoothData=Bot_BT.read();
Serial.print("Incoming from BT:");
Serial.println(BluetoothData);
}


if(flag ==1 )
call(gmotor,gnum1,gnum2); //call the respective motor for action


//Execute the functions as per the commond received through the Serial monitor or Bluetooth//
if(gmotor ==10)
```

```
left_leg_up();
if(gmotor ==11)
right_leg_up();
if(gmotor ==12)
move_left_front();
if(gmotor ==13)
move_right_front();
if(BluetoothData ==49 || gmotor ==49)
say_hi();
if(BluetoothData ==50 || gmotor ==50)
walk1();
if(BluetoothData ==51 || gmotor ==51)
walk2();
if(BluetoothData ==52 || gmotor ==52)
dance1();
if(BluetoothData ==53 || gmotor ==53)
dance2();
if(BluetoothData ==54 || gmotor ==54)
{test();test();test();}
//End of executions//
gmotor=0; //To prevet repetetion
BluetoothData = 0; //To prevet repetetion
//stay_put(); //bring the Bot to initial posotion if re-
quired
}
/
*_____
_____*/
```

// Code for Bot Functions......

```
//***Function to lift the left leg**//
void stay_put()
{
  servo5.attach(10);
  servo1.write(110);
  servo2.write(100);
  servo4.write(90);
  servo5.write(80);
  delay(20);
}
//**_____End of Function_____**//
//***Function to lift the left lef**//
void left_leg_up()
{
  Serial.println("left leg up");
  poss1 = 80;
  poss2 = 110;
  do{
  servo5.write(poss1);
  servo4.write(poss2);
  poss1++;
  poss2++;
  delay(20);
  }while(poss1 < 100 || poss2 < 140);
  call(4,130,100);
}
//**_____End of Function_____**//
//***Function to lift the left lef""//
void right_leg_up()
```

```
{
Serial.println("right leg up");
  poss1 = 80;
  poss2 = 100;
  do{
  servo4.write(poss2);
  servo5.write(poss1);
  poss1--;
  poss2--;
  delay(20);
  }while(poss1 > 50 || poss2 > 60);
  call(5,50,80);
}
//**_____End of Function_____**//
//***Function to lift the left lef**//
void move_left_front()
{
  Serial.println("moving left front");
  poss1=120;poss2=110;poss3=110;
    do{
  servo2.write(poss1);
  servo1.write(poss2);
  servo5.write(poss3);
  poss1--;
  poss2--;
  poss3--;
  delay(20);
  }while(poss1 > 100 || poss2 > 80 || poss3 > 80 );
}
//**_____End of Function_____**//
```

```
//***Function to lift the left lef**//
void move_right_front()
{
  poss1=80;poss2=100;poss3=60;
   do{
  servo1.write(poss1);
  servo2.write(poss2);
  servo4.write(poss3);
  poss1++;
  poss2++;
  poss3++;
  delay(20);
  }while(poss1<110 || poss2<120 || poss3<90);
}
//**_____End of Function_____**//
//***Function to lift the left lef**//
void say_hi()
{
 stay_put();
 right_leg_up();
 call(5,80,50); //wave up
 call(5,50,80); //wave down
 call(5,80,50); //wave up
 call(5,50,80); //wave down
 stay_put();
}
//**_____End of Function_____**//
//***Function to lift the left lef**//
void walk1()
{
```

```
stay_put();
char temp = 10; //number of steps to make * 2
do{
right_leg_up();
move_right_front();
left_leg_up();
move_left_front();
temp--;
}while(temp>0);
}
//**_____End of Function_____**//
//***Function to lift the left lef**//
void walk2()
{
stay_put();
char temp = 10; //number of steps to make * 2
do{
move_right_front();
move_left_front();
temp--;
}while(temp>0);
}
//**_____End of Function_____**//
//***Function to lift the left lef**//
void dance1()
{
stay_put();
char temp = 3; //number of steps to make * 2
do{
 poss1 = 80;
```

```
poss2 = 60;
do{
servo1.write(poss1);
servo2.write(poss2);
poss1++;
poss2++;
delay(20);
}while(poss1 <140 || poss2<120);


  poss1 = 140;
poss2 = 120;
do{
servo1.write(poss1);
servo2.write(poss2);
poss1--;
poss2--;
delay(20);
}while(poss1 >80 || poss2>60);
temp--;
}while(temp>0);
stay_put();
}
//**_____End of Function_____**//
//***Function to lift the left lef**//
void dance2()
{
stay_put();
char temp=3; //number of steps to make * 2
do{
right_leg_up(); right_leg_up();
```

```
 stay_put();
 left_leg_up();left_leg_up();
 stay_put();
 temp--;
 }while(temp>0);
 stay_put();
}
//**_____End of Function_____**//
//***Function to lift the left lef**//
void test()
{
  poss1 = 40;
  poss2 = 130;
  do{
  servo5.write(poss1);
  servo4.write(poss2);
  poss1++;
  poss2--;
  delay(5);
  }while(poss1 <120 || poss2>50);
  poss1 = 120;
  poss2 = 50;
  do{
  servo5.write(poss1);
  servo4.write(poss2);
  poss1--;
  poss2++;
  delay(5);
  }while(poss1 >40 || poss2<130);
}
```

//**_____End of Function_____**//

7. COMPUTER GENERATED REALITY UTILIZING ARDUINO AND PROCESSING

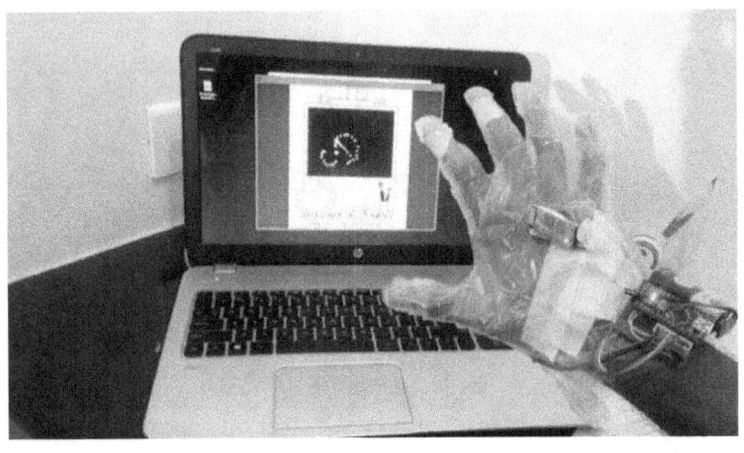

This is an extremely intriguing undertaking with regards to which we will figure out how to actualize computer generated reality utilizing Arduino and Processing. For the greater part of us, the motion picture Iron man by Jon Favreau has consistently been a motivation to assemble new things that will make our life simple and increasingly fun. I have by and by respected the Techs that are appeared in the motion picture and have for a long while been itching to fabricate something like that. Along these lines, in this undertaking I have attempted to mirror the Virtual reality stuffs that occur in the motion picture, similar to we can essentially wave our submit front of the PC and move the pointer to the ideal area and play out certain assignments.

Here I will give you how you can essentially wave your turn before webcam and draw something on your PC. I will likewise give you how you can flip lights by practically moving your hand and making clicks with your fingers noticeable all around.

Concept:

To get this going we need to use the intensity of Arduino and Processing joined. The larger part would be acquainted with Arduino, however Processing may be new for you. Preparing is an application simply like Arduino and it is additionally Open source and allowed to download. Utilizing Processing you can make straightforward framework applications, Android applications and substantially more. It likewise can do Image Processing and Voice acknowledgment. It is much the same as Arduino and is a plenty of simple to adapt, yet don't stress on the off chance that you are totally new to handling since I have composed this instructional exercise genuinely basic so anybody with intrigue can make this working in the blink of an eye.

In this instructional exercise we are utilizing Processing to make a basic System application which gives us a UI and track the situation of our hand utilizing Image handling. Presently, we need to make left snap and right snap utilizing our fingers. To get that going I have utilized two lobby sensors (one on my forefinger and the other on center finger) which will be perused by the Arduino Nano. The Arduino likewise transmits the snap status to the Computer remotely by means of Bluetooth.

It may sound confused in any case, Trust me; it isn't as hard as it sounds. So let us investigate the materials required for this undertaking to be fully operational.

Materials Required:

- Arduino Nano

- A little bit of magnet

- Lobby sensor (A3144) - 2Nos

- 9V battery

- Bluetooth Module (HC-05/HC-06)

- A couple of gloves

- Interfacing Wires Dot board.

- Handling IDE(Software)

- Arduino IDE (Software)

- A Computer with Webcam and Bluetooth (you can likewise utilize outside Bluetooth or Webcam for your PC)

Schematics and Hardware:

The equipment some portion of this undertaking is extremely basic and simple to manufacture. The total schematic is demonstrated as follows.

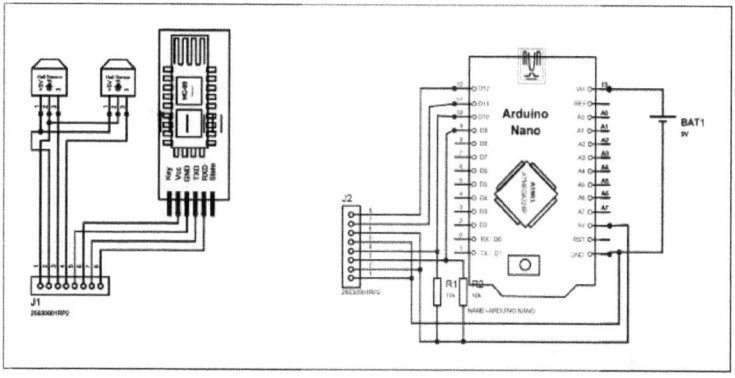

The Arduino, resistors and the berg stick pins are welded onto a spot board as demonstrated as follows.

The corridor sensor and the Bluetooth module is patched to a connector wire as demonstrated as follows.

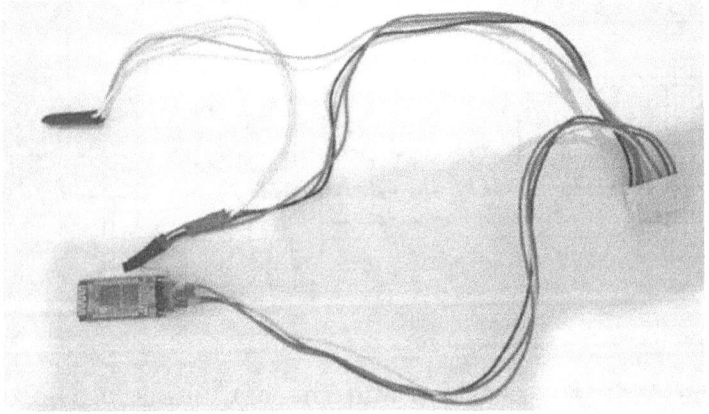

When these two areas are prepared it very well may be amassed onto gloves so it is anything but difficult to utilize. I have utilized expendable plastic gloves which can be obtained from any therapeutic shop close to you. You should ensure that the magnet goes ahead your thumb finger and the lobby sensor 1 and corridor sensor 2 ought to be available before your list and center finger separately. I have utilized duck tapes to verify the segments set up. When the segments are gathered it should look something like this.

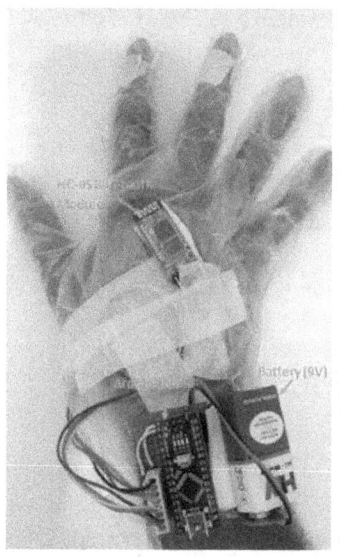 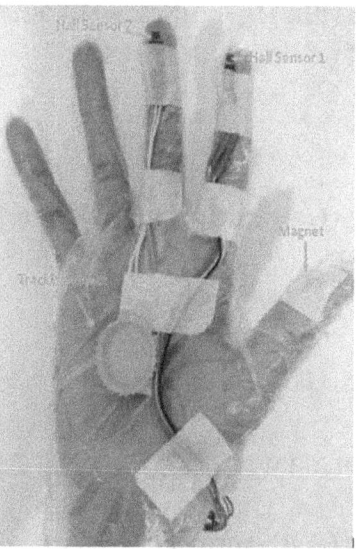

Presently let us open the Arduino IDE as well as start programming.

Program for Arduino:

The reason for this Arduino code is it to peruse the

status of the corridor sensor and communicate them utilizing the Bluetooth module. It ought to likewise get information from Bluetooth and flip the locally available LED dependent on the approaching worth. The total program is given toward the finish of this instructional exercise; I have additionally clarified not many lines underneath.

```
if (Phs1!=HallState_1 || Phs2!=HallState_2) //Check
if new keys are pressed

{

if (HallState_1==LOW && HallState_2==LOW)

Aisha.write(1);

if (HallState_1==HIGH && HallState_2==LOW)

Aisha.write(2);

if (HallState_1==LOW && HallState_2==HIGH)

Aisha.write(3);

if (HallState_1==HIGH && HallState_2==HIGH)

Aisha.write(4);
```

```
}
```

As appeared in the above lines dependent on the status of the corridor sensor the Bluetooth will compose a specific worth. For instance in the event that corridor sensor 1 is high and lobby sensor 2 is low, at that point we will communicate the vale "2" by means of the Bluetooth module. Ensure you compose the qualities to the BT module and not print them. Since it will be anything but difficult to peruse the main on Processing side just in case they are composed. Likewise the worth will possibly send in the event that it isn't as same as the past worth.

```
if(BluetoothData=='y')

digitalWrite(ledpin,HIGH);

if(BluetoothData=='n')

digitalWrite(ledpin,LOW);
```

These lines are utilized to flip the installed LED which is associated with the Pin 13, in view of the worth get by the BT module. For instance on the off chance that the module gets a 'y' at that point the LED is turned on and in the event that it gets a 'n' at that point it is killed.

Program for processing:

The motivation behind the Processing project is to make a framework application which can go about as a UI (User interface) and furthermore perform picture preparing to follow a specific article. For this situation we track the blue item that we adhered to our gloves above. The program essentially has four screens.

- Adjustment Screen

- Principle Screen

- Paint Screen

- Driven switch Screen

We can explore starting with one screen then onto the next by essentially waving our hands and hauling screens on air. We can likewise make taps on wanted spots to flip LED or even draw something on screen.

You can duplicate glue the total Processing program (given toward the end) and change it dependent on your imagination or straightforward download the EXE documents from here, and pursue the accompanying strides to dispatch the application.

- Introduce JAVA in your PC on the off chance that you have not introduced it previously

- Introduce You Cam impeccable on your PC

- Catalyst your Arduino and pair your Computer with the Bluetooth Module

- Dispatch the application record

On the off chance that everything goes fine you ought to have the option to see the LED on your Bluetooth module getting steady as well as your webcam light going ON. In the event that you have any issues contact me through the remark area and I will enable you to out.

In case you need to alter the code and incorporate more highlights with this then you can utilize the accompanying experiences of the program

The preparing IDE can be installed from here. In case you need to get familiar with preparing and make all the more fascinating undertakings then you can visit the instructional exercises here.

Handling can peruse Serial information, in this program the sequential information is originates from the Bluetooth COM port. You need to choose which COM port your Bluetooth is associate with by utilizing this line beneath

```
port = new Serial(this,Serial.list()[1],9600);
```

Here I have chosen my first COM port which is COM5 for my situation (see picture underneath) and I have referenced that by Bluetooth module keeps running on 9600 baudrate.

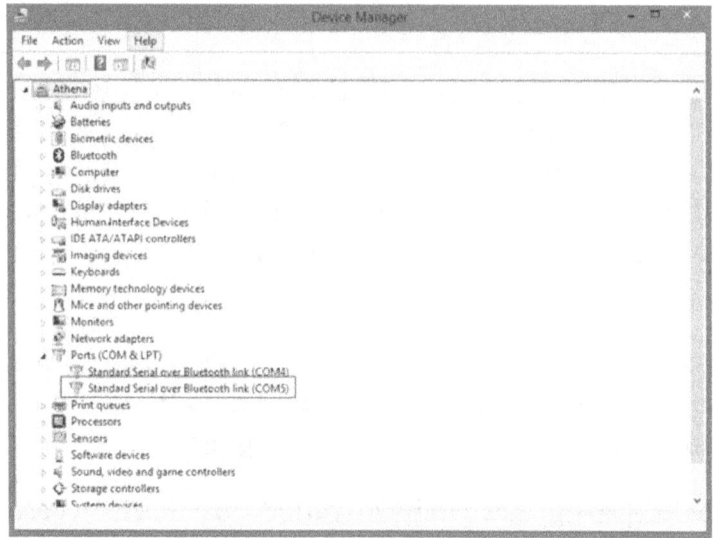

As said before handling likewise can do picture preparing, in this instructional exercise the pictures are sent inside the sketch utilizing a webcam. In each picture we track for a specific article. To find out about this you can visit this instructional exercise.

I have attempted my best to clarify the program (given toward the end) through the remark lines. You can download the records here.

Working:

When the Hardware and programming is prepared, wear the gloves and prepare for some activity. Presently, just power the Arduino and afterward dispatch the Application. The drove on the Bluetooth module should go stable. Presently it implies that your System application has built up a Bluetooth connect with your Arduino.

You will get the accompanying screen where you have to choose the article to be followed. This following can be essentially done by tapping on the item. For this situation the item is the Blue circle. Presently you can move your article and notice that the pointer pursues your item. Utilize a kind shading object and a splendid space for best outcomes.

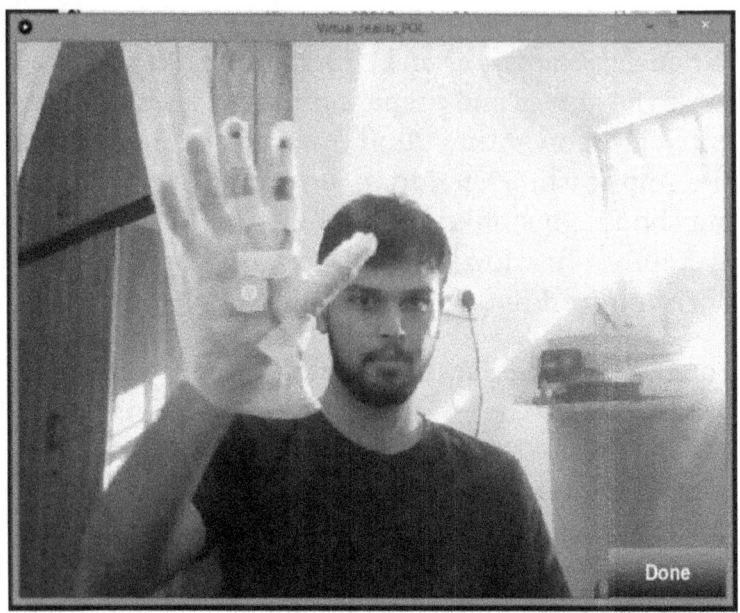

Presently contact your thumb finger with forefinger and you should see the message "Key 1 Pressed" and the when you press your thumb with center finger you should see "Key 2 Pressed" this shows everything works fine and the adjustment is finished. Presently click on the Done fasten.

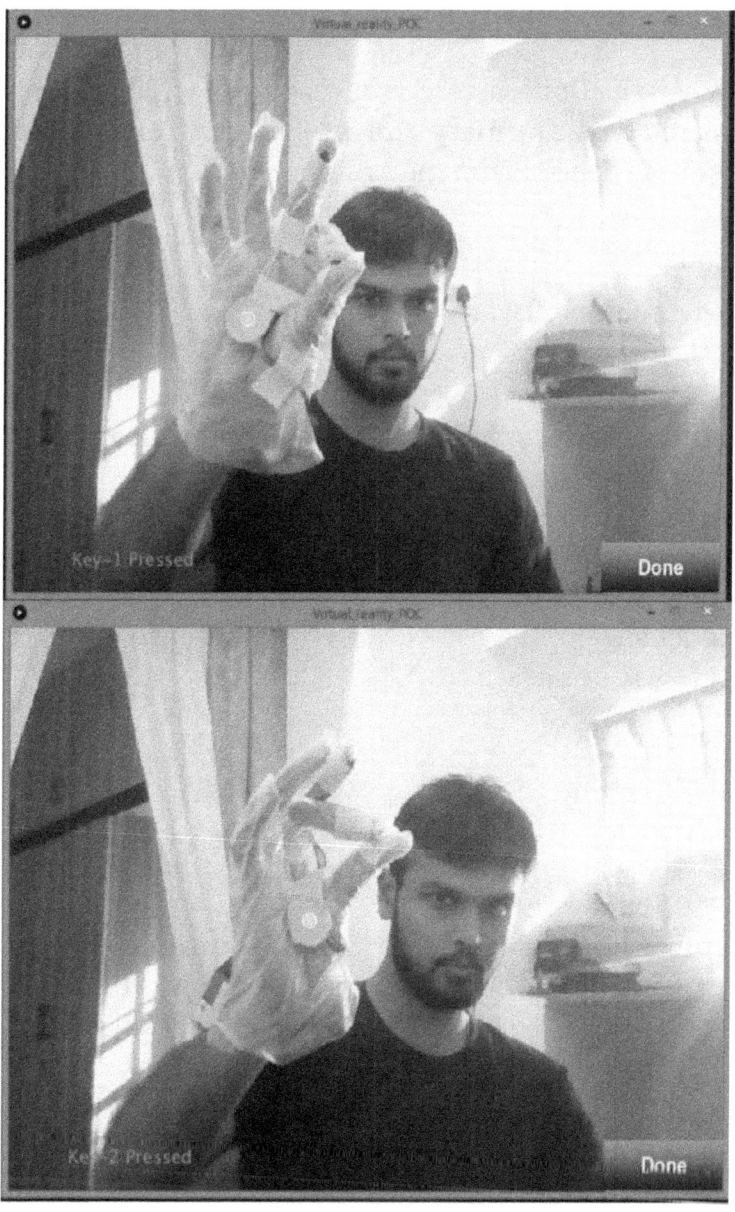

When the Done fasten is squeezed you will be coordinated to the primary screen where you can paint on air or flip the LED on the Arduino Board.

Code

```
#include <SoftwareSerial.h>// import the serial library
SoftwareSerial Aisha(11, 12); // TX, RX
int ledpin=13; //led on D13 will show blink on / off
int hall_1=9;
int hall_2=10;
int BluetoothData; // the data given from Computer
int HallState_1,HallState_2;
int change;
int Phs1,Phs2;


void setup()
{
    Aisha.begin(9600); //Bluetooth Module works at 9600 baudrate
  pinMode(ledpin,OUTPUT); //led pin as output
  pinMode(hall_1,INPUT); //hall sensor 1 as input
  pinMode(hall_2,INPUT); //hall snesor 2 is also input
}


void loop()
{
```

```
if(Aisha.available()) //if data is sent from laptop
BluetoothData=Aisha.read(); //read it and store it in
BluetoothData
Phs1=HallState_1;
Phs2=HallState_2;
HallState_1 = digitalRead(hall_1);
HallState_2 = digitalRead(hall_2);




if (Phs1!=HallState_1 || Phs2!=HallState_2) //Check
if new keys are pressed
{
if(HallState_1==LOW && HallState_2==LOW)
Aisha.write(1);

if(HallState_1==HIGH && HallState_2==LOW)
Aisha.write(2);

if(HallState_1==LOW && HallState_2==HIGH)
Aisha.write(3);

if(HallState_1==HIGH && HallState_2==HIGH)
Aisha.write(4);
}
if(BluetoothData=='y')
digitalWrite(ledpin,HIGH);
if(BluetoothData=='n')
digitalWrite(ledpin,LOW);


}
```

```
//----------- Arduino code ends---------------//
//------------Processing code starts-----------//
import processing.video.*; // Import Librarey to use
video
import processing.serial.*; //Import Librarey to use
Serial Port (Bluetooth)
//**Global Variable Declarations**//
Serial port; //port is an object variable for Serial com-
munication
int data;
boolean calibration= false;
int mirror =0;
int mirrorn =-1;
PImage        Done,Aisha,Paint,LED_Toggle,LED_on-
,LED_off;
boolean      key1,key2,key3,movePaint,PaintScreen,
PaintScreenClear,moveLED,LEDscreen;
float Paintx,Painty,avgX,avgY,LEDx,LEDy;
int count;
PImage img = createImage(380, 290, RGB);
int Px,Py;
Capture video; //create an object named video
color trackColor;  //variable to store the color that
we are going to track
float threshold = 50 ; //can be varied by the user
//_____End of variable declaration_____//
//*Function to load all the images from data folder of
the sketch*//
void loadImages()
{
```

```
 Done = loadImage("Done.png");
 Aisha = loadImage ("Aisha.png");
 Paint = loadImage("Paint.png");
 LED_Toggle = loadImage("LED_Toggle.png");
 LED_on = loadImage("LED_on.png");
 LED_off = loadImage ("LED_off.png");
}
//_____End of variable declaration_____//
//**Executes only ones**//
void setup() {
 size(800, 600);
 loadImages();
 String[] cameras = Capture.list();
 printArray(cameras);
 video = new Capture(this, cameras[34]);
 video.start();
 key1=key2=key3=false;
 Paintx=width/10;
 Painty=height/8.5;
 LEDx=width/1.1;
 LEDy=height/8.5;
    movePaint=PaintScreen=PaintScreenClear=mov-
eLED=LEDscreen=false;
 port = new Serial(this,Serial.list()[1],9600);
 println(Serial.list());
}
//**End of Setup**//
//**Triggered to update each frame of the video**//
void captureEvent(Capture video)  //when a new
image comes in
```

```
{ video.read(); } //reas it as a video
//*Function to point which color to Track*//
void Calibrate()
{
  image(video,0,0);
  imageMode(CORNERS);
   image(Done,width/1.2,height/1.1,width,height); //
position of the Done button
   if (mouseX>width/1.2 && mouseY>height/1.1) //If
mouse is within the Done button
  {
  calibration=true;
  cursor(HAND);
  mirrorn=1;
  mirror=width;
  }
fill(#1B96E0);
textSize(20);
if(key1==true) //if hall sensor 1 is active on Arduino
text("Key-1  Pressed",width/12,height/1.05); //Text
and its position
if(key2==true) //if hall sensor 2 is active on Arduino
text("Key-2  Pressed",width/12,height/1.05); //Text
and its position
}
//_____End of Calibration_____//
//*Function to represent the main Screen*//
void UI()
{
  imageMode(CORNERS);
```

```
image(Aisha,0,0,width,height);
imageMode(CENTER);


  if  ((avgX<(width/10+((width/4)/2))  && av-
gY<(height/8.5+((height/4)/2)) && key1==true) ||
(movePaint==true&&key1==true)) //if clicked in-
side the image
  {
  movePaint=true;
   image (Paint, avgX,avgY,width/4, height/4); //Drag
the image
  }
  else if(movePaint==false)
   image (Paint, Paintx,Painty,width/4, height/4); //
place the image at corner
  else
   PaintScreen=true;


  if  ((avgX>(width/1.1-((width/4)/2))  && av-
gY<(height/8.5+((height/4)/2)) && key1==true) ||
(moveLED==true&&key1==true)) //if clicked inside
the image
  {
  moveLED=true;
        image (LED_Toggle,  avgX,avgY,width/4,
height/4); //Drag the image
  }
  else if(moveLED==false)
        image (LED_Toggle,  LEDx,LEDy,width/4,
```

```
height/4); //place the image at corner
  else
   LEDscreen=true;
}
//_____End of main screen function_____//
//*Function to represent the Paint Screen*//
void Paintfun()
{
  imageMode(CENTER);
  background(#0B196A);
  image (Paint, width/2,height/2,width/1.5, height);


img.loadPixels();
for (int IX = 210, Px=0; IX<=590; IX++, Px++)
{
for (int IY = 85, Py=0; IY<=375; IY++, Py++)
 {
 if((dist(avgX,avgY,IX,IY)<4) && key1==true)
  img.pixels[(Px+(Py*img.width))] = color(255);   //
color of the paint background updated
 if(key2==true)
 PaintScreen = false;
 }
}
img.updatePixels();
image(img, width/2, height/2.6);
}
//_____End of main Paintscreen function_____//
//*Function to display Toggle LED screen*//
```

```
void LEDfun()
{
 imageMode(CENTER);
 background(255);
    image(LED_on,(width/2 - width/4), height/3,
width/4, height/5);
    image(LED_off,(width/2 + width/4), height/3,
width/4, height/5);
 textSize(50);
 textAlign(CENTER);
  if (key1==true && avgX<300 && avgY>150 &&
avgX>95 && avgY<260)
 {fill(#751EE8);
 text("LED turned on",width/2,height/1.5);
 port.write(121);
 }
  if (key1==true && avgX<700 && avgY>150 &&
avgX>500 && avgY<260)
 {fill(#FC0808);
 text("LED turned  off",width/2,height/1.5);
 port.write(110);
 }
}
//_____End of main LEDscreen function_____//
//*Function to know which key is pressed*//
void key_select() {


 switch(data){
 case 1:
  key1=true; key2=true;
```

```
  break;


  case 2:
   key1=false; key2=true;
   break;


  case 3:
   key1=true; key2=false;
   break;


  case 4:
   key1=false; key2=false;
   break;
}
}
//_____End of function_____//
void draw(){
  if (port.available()>0) //if there is an incoming BT
value
  {
   data=port.read(); //read the BT incoming value and
save in data
   println(key1,key2,data); //print for debugging
   key_select(); //toggle the variable key 1 and key2
  }


  video.loadPixels();
```

```
if (calibration==false) //no calibration done
Calibrate(); //Calibrate Screen
if (calibration==true && (PaintScreen==false || LED-
screen==false))
UI(); //Main Screen
if (PaintScreen==true && calibration ==true)
Paintfun(); //Paint Screen
if (LEDscreen==true && calibration ==true)
LEDfun(); //LED toffle screen


if (key2==true)
    movePaint=PaintScreen=PaintScreenClear=mov-
eLED=LEDscreen=false; //go back to main screen

avgX = avgY = count = 0;
// Begin loop to walk through every pixel
for (int x = 0; x < video.width; x++) {
 for (int y = 0; y < video.height; y++) {
  int loc = x + y * video.width;
  // What is current color
  color currentColor = video.pixels[loc];
  float r1 = red(currentColor);
  float g1 = green(currentColor);
  float b1 = blue(currentColor);
  float r2 = red(trackColor);
  float g2 = green(trackColor);
  float b2 = blue(trackColor);
  float d = distSq(r1, g1, b1, r2, g2, b2);
  if (d < threshold*threshold) {
```

```
    stroke(255);
    strokeWeight(1);
    // point((mirror-x)*mirrorn, y);
    avgX += x;
    avgY += y;
    count++;
  }
 }
}
if(count > 0){
 avgX = avgX / count;
 avgY = avgY / count;
 // Draw a circle at the tracked pixel
 fill(#21FADB);
 avgX = (mirror-avgX)*mirrorn;
 ellipse(avgX, avgY, 15, 15);
 }
}
float distSq(float x1, float y1, float z1, float x2, float
y2, float z2){
   float d = (x2-x1)*(x2-x1) + (y2-y1)*(y2-y1) +(z2-
z1)*(z2-z1);
 return d;
}
void mousePressed(){
 if(calibration==false)
 {
 int loc = mouseX + mouseY*video.width;
 trackColor = video.pixels[loc]; //load the color to be
tracked
```

```
  }
}
```

8. ARDUINO BASED VEHICLE ACCIDENT ALERT SYSTEM UTILIZING GPS, GSM AND ACCELEROMETER

In our past instructional exercises, we have found out about How to interface Global Positioning System module with Computer, how to fabricate an Arduino GPS Clock and how to Track vehicle utilizing GSM and GPS. Here in this undertaking, we are gonna to assemble an Arduino based vehicle mishap ready framework utilizing GPS, GSM and accelerometer. Accelerometer identifies the abrupt change in the tomahawks of vehicle and GSM module sends the alarm message on your Mobile Phone with the area of the mishap. Area of mishap is sent as Google Map interface, got from the scope and longitude from GPS module. The Message additionally contains the speed of vehicle in tangles. This Vehicle Accident ready task can likewise be utilized as a Tracking System and significantly more, by simply rolling out not many improvements in equipment and programming.

Components Required:

- Arduino Uno

- GPS Module (SIM28ML)
- GSM Module (SIM900A)
- 16x2 LCD
- Accelerometer (ADXL335)
- Connecting Wires
- Power Supply
- Breadboard or PCB
- 10 K-POT
- Power supply 12v 1amp

Prior to going into Project, we will talk about GPS, GSM as well as Accelerometer.

GPS Module and Its Working:

GPS represents Global Positioning System and used to recognize the Latitude and Longitude of any area on the Earth, with careful UTC time (Universal Time Co-ordinated). GPS module is utilized to follow the area of mishap in our undertaking. This gadget gets the directions from the satellite for every single second, with time and date. We have recently removed $GP-GGA string in Vehicle Tracking System to discover the Latitude and Longitude Coordinates.

GPS module sends the information identified with following situation progressively, and it sends such huge numbers of information in NMEA position (see the screen capture beneath). NMEA organization comprises a few sentences, where we just need one sentence. This sentence begins from $GPGGA and contains the directions, time and other helpful data. This GPGGA is alluded to Global Positioning System Fix Data. Find out about NMEA sentences and perusing GPS information here.

We can concentrate organize from $GPGGA string by including the commas in the string. Assume you discover $GPGGA string and stores it in an exhibit, at that point Latitude can be found after two commas and Longitude can be found after four commas. Presently, this scope and longitude can be placed in different exhibits.

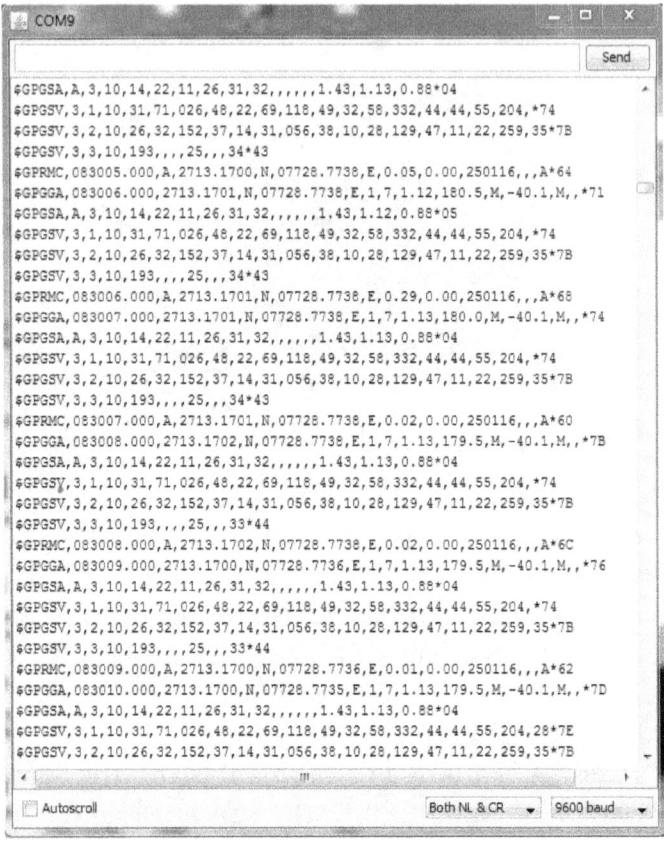

The following is the $GPGGA String, alongside its portrayal:

**$GPG-
GA,104534.000,7791.0381,N,06727.4434,E,1,08,
0.9,510.4,M,43.9,M,,*47**

$GPGGA,HHMMSS.SSS,latitude,N,longi-tude,E,FQ,NOS,HDP,altitude,M,height,M,,check-sum data

Identifier	Description
$GPGGA	Global Positioning system fix data
HHMMSS.SSS	Time in hour minute seconds and milliseconds format.
Latitude	Latitude (Coordinate)
N	Direction N=North, S=South
Longitude	Longitude(Coordinate)
E	Direction E= East, W=West
FQ	Fix Quality Data
NOS	No. of Satellites being Used
HDP	Horizontal Dilution of Precision
Altitude	Altitude (meters above from sea level)
M	Meter
Height	Height
Checksum	Checksum Data

GSM Module:

The SIM900 is a finished Quad-band GSM/GPRS Module which can be inserted effectively utilized by client or specialist. SIM900 GSM Module gives an industry-standard interface. SIM900 conveys Global System for Mobile/General Packet Radio Service 850/900/1800/1900MHz execution for voice, SMS, Data with low power usage. It is effectively accessible in the market.

- SIM900 planned by utilizing single-chip processor coordinating AMR926EJ-S center

- Quad - band Global System for Mobile/General Packet Radio Service module in little size.

- General Packet Radio Service Enabled

AT Command:

AT implies ATTENTION. This order is utilized to control GSM module. There are a few directions for calling and informing that we have utilized in a significant number of our past GSM ventures with Arduino. For testing GSM Module we utilized AT direction. In the wake of getting AT Command GSM Module react with OK. It implies GSM module is working fine. The following is some AT directions we utilized here in this task:

ATE0 For echo off

AT+CNMI=2,2,0,0,0 <ENTER>   Auto opened message Receiving. (No need to open message)

ATD<Mobile Number>; <ENTER>   making a call (ATD+919610126059;\r\n)

AT+CMGF=1 <ENTER>   Selecting Text mode

AT+CMGS="Mobile Number" <ENTER> Assigning recipient's mobile number

>>Now we can write our message

>>After writing message

Ctrl+Z send message command (26 in decimal).

ENTER=0x0d in HEX

(To study GSM module, Check our different GSM ventures with different microcontrollers here)

Accelerometer:

Stick Description of accelerometer:

- Vcc 5 volt supply ought to associate at this stick.

- X-OUT This stick gives an Analog yield in x course

- Y-OUT This stick give an Analog Output in y course

- Z-OUT This stick gives an Analog Output in z heading

- GND Ground

- ST This stick utilized for set affectability of sensor

Likewise check our different undertakings utilizing Accelerometer: Ping Pong Game utilizing Arduino and Accelerometer Based Hand Gesture Controlled Robot.

Circuit Explanation:

Circuit Connections of this Vehicle Accident Alert System Project is straightforward. Here Tx stick of GPS module is straightforwardly associated with advanced stick number 10 of Arduino. By utilizing Software Serial Library here, we have permitted sequential correspondence on stick 10 and 11, and made them Rx and Tx separately and left the Rx stick of GPS Module open. As a matter of course Pin 0 and 1 of Arduino are utilized for sequential correspondence yet by utilizing the SoftwareSerial library, we can permit sequential correspondence on other advanced pins of the Arduino. 12 Volt supply is utilized to control the GPS Module.

GSM module's Tx as well as Rx pins of are legitimately associated with stick D2 and D3 of Arduino. For GSM interfacing, here we have additionally utilized programming sequential library. GSM module is likewise controlled by 12v inventory. A discretionary LCD's information pins D4, D5, D6, as well as D7 are associated with stick number 6, 7, 8, as well as 9 of Arduino. Order stick RS and EN of LCD are associated with stick number 4 and 5 of Arduino and RW stick is legitimately associated with ground. A Potentiometer is additionally utilized for setting differentiation or splendor of LCD.

An Accelerometer is included this framework for identifying a mishap and its x,y, and z-pivot ADC yield pins are legitimately associated with Arduino ADC stick A1, A2, and A3.

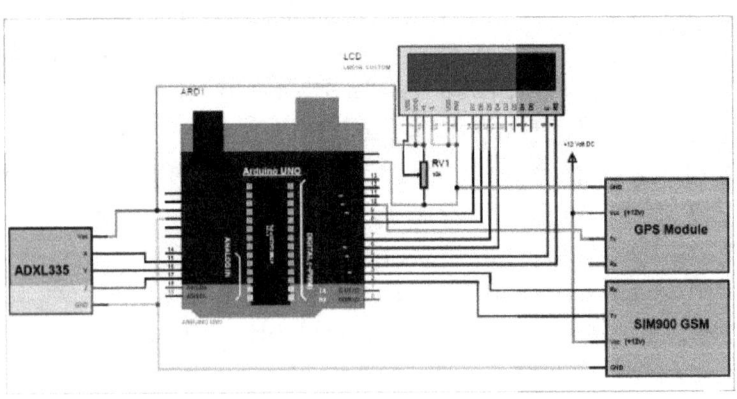

Working Explanation:

In this undertaking, Arduino is utilized for controlling entire the procedure with a GPS Receiver and GSM module. GPS Receiver is used for recognizing directions of the vehicle, Global System for Mobile module is used for sending the ready SMS with the directions as well as the connection to Google Map. Accelerometer to be specific ADXL335 is utilized for distinguishing mishap or abrupt change in any hub. What's more, a discretionary 16x2 LCD is likewise utilized for showing status messages or organizes. We have utilized GPS Module SIM28ML and GSM Module SIM900A.

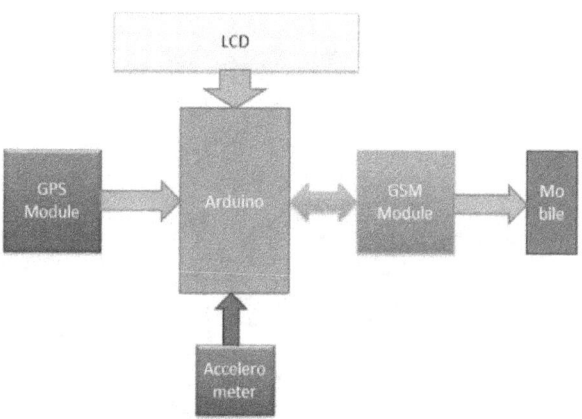

At the point when we are prepared with our equipment in the wake of programming, we can introduce it in our vehicle and power it up. Presently at whatever point there is a mishap, the vehicle gets tilt and accelerometer changes his hub esteems. These qualities read by Arduino and checks if any change happens in any hub. In the event that any change happens, at that point Arduino peruses organizes by separating $GPGGA String from GPS module information (GPS working clarified above) and send SMS to the predefined number to the police or rescue vehicle or relative with the area directions of mishap place. The message likewise contains a Google Map connect to the mishap area, with the goal that area can be effectively followed. At the point when we get the message then we just need to tap the connection and we will divert to the Google guide and afterward we can see the definite area of the vehicle. Speed of Vehicle, in tangles (1.852 KPH), is likewise sent in the SMS and showed on the LCD board.

Here in this task, we can set the affectability of Accelerometer by putting min and max esteem in the code.

Here in the demo have utilized given qualities:

```
#define minVal -50

#define MaxVal 50
```

In any case, for better outcomes you can utilize 200 instead of 50, or can set by your prerequisite.

Programming Explanation:

Complete Program has been given underneath in Code area; here we are clarifying its different capacities in a nutshell.

First we have incorporated all the necessary libraries or headers records and announced different factors for counts and putting away information transitory.

After this, we have made a capacity void initModule(String cmd, singe *res, int t) to introduce the GSM module and checking its reaction utilizing AT directions.

```
void initModule(String cmd, char *res, int t)

{
```

```
while(1)

{

Serial.println(cmd);

Serial1.println(cmd);

delay(100);

while(Serial1.available()>0)

{

  if(Serial1.find(res))

  {

  Serial.println(res);

  delay(t);

  return;

  }

  else

  {
```

```
    Serial.println("Error");

   }

  }

  delay(t);

 }

}
```

After this, in void arrangement() work, we have instated equipment and programming sequential correspondence, LCD, GPS, GSM module and accelerometer.

```
void setup()

{

  Serial1.begin(9600);

  Serial.begin(9600);

  lcd.begin(16,2);

  lcd.print("Accident Alert ");
```

```
lcd.setCursor(0,1);

lcd.print(" System ");

delay(2000);

lcd.clear();

.... ......

...... .....
```

Accelerometer alignment procedure is likewise done in arrangement circle. In this, we have taken a few examples and afterward locate the normal qualities for the x-pivot, y-hub, and z-hub. What's more, store them in a variable. At that point we have utilized these example esteems to peruse changes in accelerometer pivot when vehicle gets tilt (mishap).

```
lcd.print("Callibrating");

lcd.setCursor(0,1);

lcd.print("Acceleromiter");

for(int i=0;i<samples;i++)

{
```

```
  xsample+=analogRead(x);

  ysample+=analogRead(y);

  zsample+=analogRead(z);

}

xsample/=samples;

ysample/=samples;

zsample/=samples;

Serial.println(xsample);

Serial.println(ysample);

Serial.println(zsample);
```

After this, in the void circle() work, we have perused accelerometer pivot esteems and done an estimation to concentrate changes with the assistance of tests that are taken in Calibration. Presently in case any progressions are pretty much, at that point characterized level then Arduino makes an impression on the predefined number.

```
void loop()

{

   int value1=analogRead(x);

   int value2=analogRead(y);

   int value3=analogRead(z);

   int xValue=xsample-value1;

   int yValue=ysample-value2;

   int zValue=zsample-value3;

   Serial.print("x=");

   Serial.println(xValue);

   Serial.print("y=");

   Serial.println(yValue);

   Serial.print("z=");

   Serial.println(zValue);
```

....

....

Here we have likewise made some other capacity for different puposes like void gpsEvent() to get GPS facilitates, void coordinate2dec() for separating organizes from the GPS string and convert them into Decimal qualities, void show_coordinate() for showing esteems over sequential screen and LCD, lastly the void Send() for sending ready SMS to the predefined number.

Complete code is given underneath, you can check every one of the capacities in the code.

Code

```
#include<SoftwareSerial.h>
SoftwareSerial Serial1(2,3); //make RX arduino line is pin 2, make TX arduino line is pin 3.
SoftwareSerial gps(10,11);
#include<LiquidCrystal.h>
LiquidCrystal lcd(4,5,6,7,8,9);
#define x A1
#define y A2
#define z A3
int xsample=0;
int ysample=0;
int zsample=0;
#define samples 10
```

```
#define minVal -50
#define MaxVal 50
int i=0,k=0;
int gps_status=0;
float latitude=0;
float logitude=0;
String Speed="";
String gpsString="";
char *test="$GPRMC";
void initModule(String cmd, char *res, int t)
{
 while(1)
 {
  Serial.println(cmd);
  Serial1.println(cmd);
  delay(100);
  while(Serial1.available()>0)
  {
   if(Serial1.find(res))
   {
   Serial.println(res);
   delay(t);
   return;
   }
   else
   {
   Serial.println("Error");
   }
  }
  delay(t);
```

```
  }
}
void setup()
{
 Serial1.begin(9600);
 Serial.begin(9600);
 lcd.begin(16,2);
 lcd.print("Accident Alert ");
 lcd.setCursor(0,1);
 lcd.print(" System  ");
 delay(2000);
 lcd.clear();
 lcd.print("Initializing");
 lcd.setCursor(0,1);
 lcd.print("Please Wait...");
 delay(1000);


  Serial.println("Initializing....");
 initModule("AT","OK",1000);
 initModule("ATE1","OK",1000);
 initModule("AT+CPIN?","READY",1000);
 initModule("AT+CMGF=1","OK",1000);
 initModule("AT+CNMI=2,2,0,0,0","OK",1000);
 Serial.println("Initialized Successfully");
 lcd.clear();
 lcd.print("Initialized");
 lcd.setCursor(0,1);
 lcd.print("Successfully");
 delay(2000);
 lcd.clear();
```

```
lcd.print("Callibrating");
lcd.setCursor(0,1);
lcd.print("Acceleromiter");
for(int i=0;i<samples;i++)
{
 xsample+=analogRead(x);
 ysample+=analogRead(y);
 zsample+=analogRead(z);
}
xsample/=samples;
ysample/=samples;
zsample/=samples;

Serial.println(xsample);
Serial.println(ysample);
Serial.println(zsample);
delay(1000);


 lcd.clear();
lcd.print("Waiting For GPS");
lcd.setCursor(0,1);
lcd.print("   Signal   ");
delay(2000);
gps.begin(9600);
get_gps();
show_coordinate();
delay(2000);
lcd.clear();
lcd.print("GPS is Ready");
 delay(1000);
```

```
lcd.clear();
lcd.print("System Ready");
Serial.println("System Ready..");
}
void loop()
{
  int value1=analogRead(x);
  int value2=analogRead(y);
  int value3=analogRead(z);
  int xValue=xsample-value1;
  int yValue=ysample-value2;
  int zValue=zsample-value3;


   Serial.print("x=");
  Serial.println(xValue);
  Serial.print("y=");
  Serial.println(yValue);
  Serial.print("z=");
  Serial.println(zValue);
   if(xValue < minVal || xValue > MaxVal || yValue <
minVal || yValue > MaxVal || zValue < minVal || zValue
> MaxVal)
  {
  get_gps();
  show_coordinate();
  lcd.clear();
  lcd.print("Sending SMS ");
  Serial.println("Sending SMS");
  Send();
```

```
    Serial.println("SMS Sent");
    delay(2000);
    lcd.clear();
    lcd.print("System Ready");
   }
}
void gpsEvent()
{
 gpsString="";
 while(1)
 {
  while (gps.available()>0)      //Serial incoming data
from GPS
  {
   char inChar = (char)gps.read();
    gpsString+= inChar;              //store incoming data
from GPS to temparary string str[]
   i++;
   // Serial.print(inChar);
   if(i < 7)
   {
     if(gpsString[i-1] != test[i-1])      //check for right
string
   {
    i=0;
    gpsString="";
   }
   }
  if(inChar=='\r')
  {
```

```
  if(i>60)
  {
   gps_status=1;
   break;
  }
  else
  {
   i=0;
  }
  }
 }
 if(gps_status)
  break;
 }
}
void get_gps()
{
 lcd.clear();
 lcd.print("Getting GPS Data");
 lcd.setCursor(0,1);
 lcd.print("Please Wait.....");
 gps_status=0;
 int x=0;
 while(gps_status==0)
 {
 gpsEvent();
 int str_lenth=i;
 coordinate2dec();
 i=0;x=0;
 str_lenth=0;
```

```
  }
}
void show_coordinate()
{
  lcd.clear();
  lcd.print("Lat:");
  lcd.print(latitude);
  lcd.setCursor(0,1);
  lcd.print("Log:");
  lcd.print(logitude);
  Serial.print("Latitude:");
  Serial.println(latitude);
  Serial.print("Longitude:");
  Serial.println(logitude);
  Serial.print("Speed(in knots)=");
  Serial.println(Speed);
  delay(2000);
  lcd.clear();
  lcd.print("Speed(Knots):");
  lcd.setCursor(0,1);
  lcd.print(Speed);
}
void coordinate2dec()
{
  String lat_degree="";
  for(i=20;i<=21;i++)
    lat_degree+=gpsString[i];


  String lat_minut="";
```

```
  for(i=22;i<=28;i++)
   lat_minut+=gpsString[i];
 String log_degree="";
  for(i=32;i<=34;i++)
   log_degree+=gpsString[i];
 String log_minut="";
  for(i=35;i<=41;i++)
   log_minut+=gpsString[i];


   Speed="";
   for(i=45;i<48;i++)          //extract longitude from
string
   Speed+=gpsString[i];


   float minut= lat_minut.toFloat();
  minut=minut/60;
  float degree=lat_degree.toFloat();
  latitude=degree+minut;


   minut= log_minut.toFloat();
  minut=minut/60;
  degree=log_degree.toFloat();
  logitude=degree+minut;
}
void Send()
{
 Serial1.println("AT");
 delay(500);
```

```
 serialPrint();
 Serial1.println("AT+CMGF=1");
 delay(500);
 serialPrint();
 Serial1.print("AT+CMGS=");
 Serial1.print("");
 Serial1.print("9821757249");   //mobile no. for SMS alert
 Serial1.println("");
 delay(500);
 serialPrint();
 Serial1.print("Latitude:");
 Serial1.println(latitude);
 delay(500);
 serialPrint();
 Serial1.print("longitude:");
 Serial1.println(logitude);
 delay(500);
 serialPrint();
 Serial1.print(" Speed:");
 Serial1.print(Speed);
 Serial1.println("Knots");
 delay(500);
 serialPrint();
        Serial1.print("http://maps.google.com/maps?&z=15&mrt=yp&t=k&q=");
 Serial1.print(latitude,6);
    Serial1.print("+");                     //28.612953, 77.231545 //28.612953,77.2293563
 Serial1.print(logitude,6);
```

```
  Serial1.write(26);
  delay(2000);
  serialPrint();
}
void serialPrint()
{
 while(Serial1.available()>0)
 {
  Serial.print(Serial1.read());
 }
}
```

9. DIY SPEEDOMETER UTILIZING ARDUINO AS WELL AS PROCESSING ANDROID APP

In this task we make a Cool Speedometer for bicycles or any automotives by utilizing Arduino which communicate the speed utilizing Bluetooth to an Android application that we made utilizing Processing. The total venture is fueled by a 18650 Lithium cell and henceforth exceptionally compact alongside your vehicle. To zest it up more I have included the alternative of charging your cell phone as it shows your speed. Truly, you can likewise utilize this as a power bank for your mobiles in a hurry as the 18650 has high charge thickness and can without much of a stretch be charged and released.

I will totally direct you without any preparation till the fruition, speedometer can he snared to our vehicle and tried. The cool component here is that you can alter your android application for your personalisation and add more highlights as indicated by your imagination. In case, you would prefer not to assemble the application independent from anyone else and simply manufacture the Arduino part at that point no stresses, simply install the APK record (read further) and introduce in your Android cell phone.

So we should perceive what materials we would need to construct this task, and plan our financial limit. Every one of these segments are effectively accessible.

Hardware Requirements:

- Arduino Pro Mini (5V 16MHz)
- 3V to 5V DC-DC Boost converter with USB yield charger
- FTDI board (for programming small you can likewise utilize UNO)
- TP4056 Lithium battery Module
- Bluetooth Module (HC-05/HC-06)
- Corridor impact sensor (US1881/04E)
- 18650 Lithium Cell
- Little bit of magnets
- Perf Board
- Berg sticks connectors (Male and Female)
- Patching Kit
- Little fenced in area boxes to mount the pack.

Programming Requirements:

- Arduino IDE

- Windows/Mac PC

- Handling IDE with android ADK (Only on the off chance that you need to code your very own application.)

- Android Mobile Phone.

It may resemble a bunch of segments and materials, however trust me once you complete this undertaking you would feel they merit the time the exertion.

Measuring Speed using Hall Sensor and Arduino:

Before we get our hands on the equipment, let us know how we are really going to quantify the speed utilizing Arduino. There is bunches of approach to quantify the speed of a vehicle utilizing Arduino, yet utilizing a corridor sensor is the most financial and least demanding method for doing it. A Hall Sensor is a segment that distinguishes the extremity of a magnet. For instance at whatever point one specific shaft of the magnet is brought close to the sensor, the sensor will change its state. There are many kinds of corridor sensors accessible you can utilize any of those in this task however ensure that it is a computerized lobby sensor.

To gauge the speed we need to stick a little bit of magnet onto the wheel of the vehicle, each time the magnet crosses the corridor sensor it will recognize it and sends the data to the Arduino.

A hinder will be gotten by the Arduino each time the magnet is recognized. We run a ceaseless clock by utilizing millis() work and compute the time taken for the wheel to finish two turns (to limit blunder) by utilizing the beneath formulae:

Timetaken = millis() – pevtime;

When we realize the time taken we can compute the rpm by utilizing the beneath formulae:

```
rpm = (1000/timetaken) * 60;
```

Where (1000/timetaken) gives the rps (Revolutions every second) and it is duplicated by 60 to change over rps to rpm (Revolutions every moment).

In the wake of computing the rpm we can ascertain the speed of the vehicle utilizing the underneath formulae gave we know the sweep of the wheel.

```
v = radius_of_wheel * rpm * 0.37699;
```

The Arduino, in the wake of computing the speed, will communicate it utilizing the Bluetooth Module. The total code has been given underneath in Code area. Additionally check our different activities including Bluetooth Module HC-05 here.

Schematics and Hardware Part:

The total Circuit chart of the venture is given beneath:

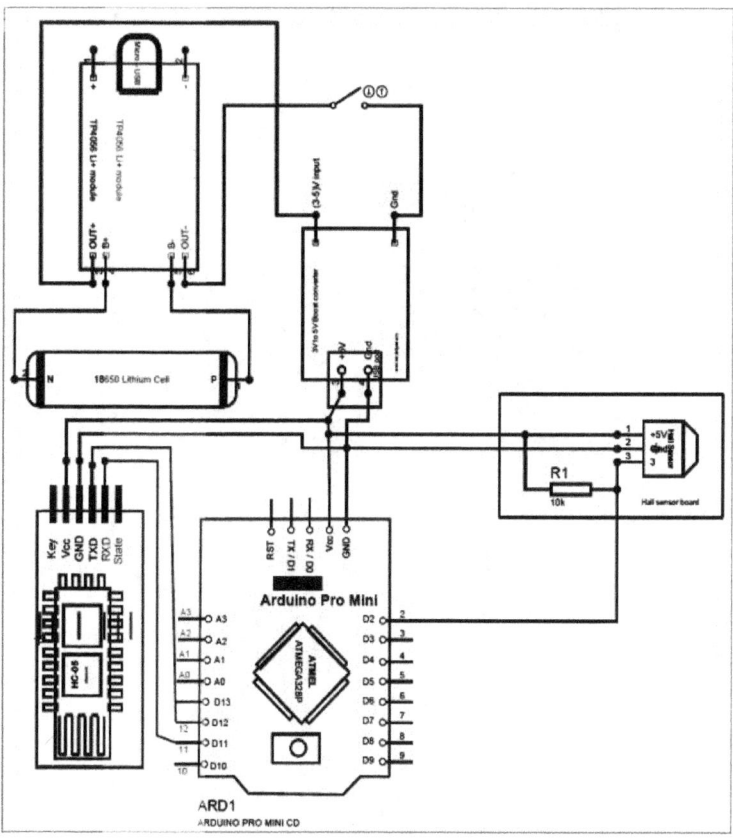

Here, the equipment part is part into two sections one is the primary board which contains all the fundamental records. The other board just comprises of a corridor sensor and a resistor which will be mounted close to the wheel. Give us a chance to begin assembling the principle board.

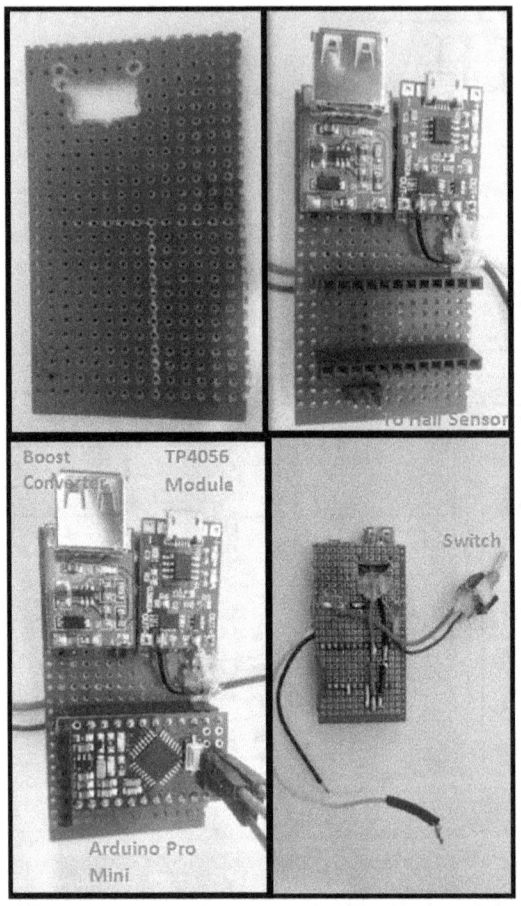

When the association is made given us a chance to test the set up by utilizing our 18650 Lithium battery. Lithium battery is exceptionally dangerous in nature, subsequently it must be taken care of with outrageous alert. It is therefore why we utilize a TP4056 Lithium Battery Charging Module. This module has over charge/release security and Reverse extremity assurance. Henceforth the battery can be effectively charged utilizing an ordinary miniaturized scale USB charger and can be securely released till it arrives at the under voltage cut beyond reach. Some significant subtleties of the TP4056 charge module is given in the table beneath.

Parameters :	Value per Cell :

Under Voltage cut-off	2.4V
Over voltage Cut-off	4.2V
Charging current	1A
Protection	Over Voltage and reverse polarity protection
IC's present	TP4056 (charger IC) and DW01 Protection IC
Indication LED's	Red- Charging in Progress Green – Charge Complete

Presently, let us start with the Hall Sensor Board. This board just contains two parts one it the 10K resistor and the other is the lobby sensor. The associations can be made as appeared in the schematics above. When the board is prepared, interface them utilizing jumper wires according to the schematics. When it is done it should look something like this.

Another urgent advance in the task is interfacing the 18650 battery to the B+ as well as B-terminals of the TP4056 module utilizing a wire. Since Li+ cells are touchy it is profoundly not prescribed to utilize a binding iron over these cells. In spite of the fact that individuals have done it, it is exceptionally dangerous and can without much of a stretch end up in a major chaos. Consequently the easy method to do it, is to utilize magnets as demonstrated as follows

Essentially bind the wire to a little bit of magnet and afterward stick the magnets to the terminals of the battery (they get pulled in to terminals well overall) as appeared previously. You may utilize some duck tap to further verify the situation of the magnet.

Programming the Arduino:

The program for this undertaking is extremely basic. We simply need to compute the speed of pivoting wheel by utilizing the corridor sensor intrude on sources of info and communicate the determined speed over the air utilizing Bluetooth Module. The total program is given in the Code area underneath and clarified utilizing the remark lines.

Each time the corridor sensor distinguishes the magnet it triggers an intrude. This interfere with capacity is called by the magnet_detect() work. This is where the rpm of the vehicle is determined.

When the rpm is determined the speed of the wheel is determined on the up and up () work. When the code is prepared lets dump it to our Arduino star small and test its working.

Android Mobile Application for Speedometer:

The Android application for this task is made utilizing programming called Processing. In case you are not keen on making your own Android application and might want to simply introduce the one utilized here, you can download the APK document and introduce it legitimately in your Smart Phone by following the beneath steps.

1. You can legitimately download the APK document from the underneath connect. This APK document is made for Android form 4.4.2 or more (Kitkat an above). Concentrate the APK record from the compress document.

Android Application for Speedometer

2. Move the .Apk record from your PC to your cell phone.

3. Empower introducing application from Unknown sources in your android settings.

4. Introduce the application.

In case that effectively introduced, you will discover

the application named "Processing_code" introduced on your telephone as demonstrated as follows:

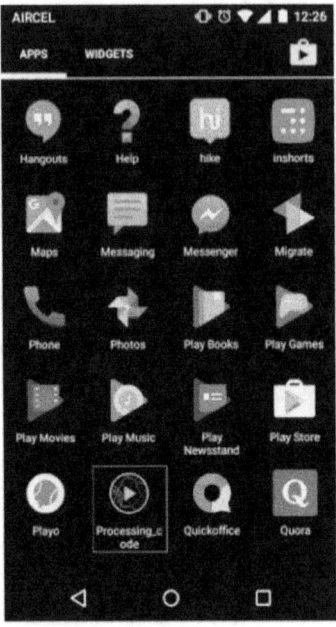

Build up your very own Application utilizing Processing:

Possibly you can utilize the .APK record given above or you can construct your very own application utilizing Processing as clarified here. You can download the all the Processing Android application code from here. The program is self-clarified utilizing the remark lines. Yet, if you have any issue or in case you need to get your application an altered a piece please utilize the remark area and I will enable you to out.

The android program builds up an association with our Bluetooth module during start up of the application and gets the speed of the vehicle which was determined and communicated by the Arduino Pro smaller than normal. I have made a little illustrations likewise to show the speed utilizing an Analog speedometer to make it look somewhat alluring. You can think of your own thoughts and change the code to customize it for your needs. Likewise check our other Processing Projects to get familiar with it:

- Ping Pong Game utilizing Arduino

- Advanced mobile phone Controlled FM Radio utilizing Processing.

- Arduino Radar System utilizing Processing and Ultrasonic Sensor

When you have introduced the application to your cell phone its opportunity to test our undertaking. Yet, we have not mounted our pack to a vehicle yet. How about we do it.

Mounting the Speedometer kit to a vehicle:

I have mounted this unit over my bi-cycle and tried it, and it brings about the ideal result. Mounting the pack is left to your innovativeness, you can get your own little box from a shop and drill gaps for the wires and associations and mounted it to your vehicle. One normal significant thing to note is that the magnet ought to be adhered to the edge of the haggle corridor sensor ought to be mounted as close as conceivable to the magnet with the goal that each time the magnet crosses the lobby sensor it ought to have the option to distinguish it, the course of action is demonstrated as follows.

Since I have a 3D printer with me, I planned my very own nooks to make them look great and in way that it tends to be effectively mounted and disengaged from our bicycle for charging the battery. So in the event that you have a 3D printer or on the off chance that you can access one to print not many materials keep perusing, else avoid this part and utilize your very own inventiveness to mount these things. Find out how to client 3D printer here.

In case you have chosen to utilize my structure documents and print your walled in areas at that point ensure your principle perf board is near the beneath measurements

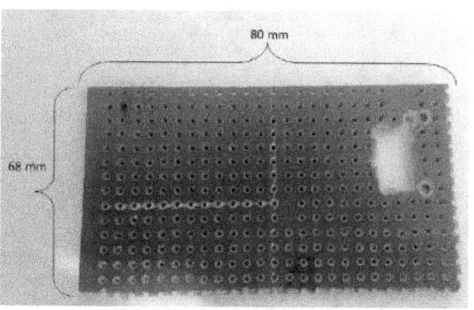

The total Design and STL records for 3D printing can be installed from here. In case the board is like what is have made here, at that point you can straightforwardly 3D print your fenced in areas utilizing the given STL records or else you can utilize the Design documents and alter it as indicated by your board.

Give us a chance to begin with the 3D printed little fenced in area which will be utilized for the corridor sensor module. Print the fenced in area, place the circuit into it and bridle your wires however the opening gave and afterward mount it to your vehicle with the goal that the corridor sensor is close to the magnet as demonstrated as follows.

It is prescribed to show the fundamental board before structuring the nook for it with the goal that we can ensure it fits in appropriately, in light of the fact that trust me it would be bad dream when you print your walled in area for 6 extended periods and at last it won't fit into your perf board. The model board for my fundamental perf board is demonstrated as follows.

Presently it will be anything but difficult to structure the primary walled in area box. I have structured the principle confine two documents, with the goal that one piece of the case will hold the hardware and the other will be forever fixed to the cycle utilizing cinches and jolts. These two sections can without much of a stretch be fixed to assemble to cause a total fenced in area and afterward to be isolated when we have to energize our lithium battery or work on our gadgets.

When the initial segment of the nook is planned and

printed lets place every one of our parts inside as appeared underneath and it should look something like this..

As should be obvious there are two openings before the container, one is utilized for the USB through which we can charge our cell phone. The other is for the small scale USB utilizing which we can charge our lithium battery.

Presently let us print the second piece of the fundamental walled in area and check in the event that it fits the initial segment true to form.

When we are happy with the parts we can mount the second piece of the fenced in area utilizing a C-brace and a few nuts and fastener as demonstrated as follows:

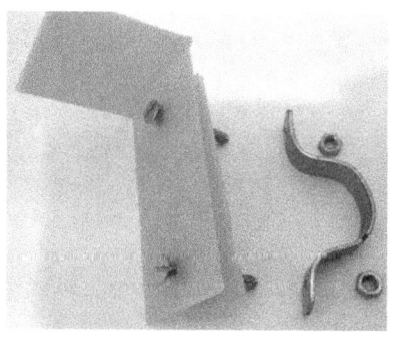

Presently let us associate the battery to our fundamental circuit utilizing magnets and tape as talked about above and keep it forever safe inside our walled in area.

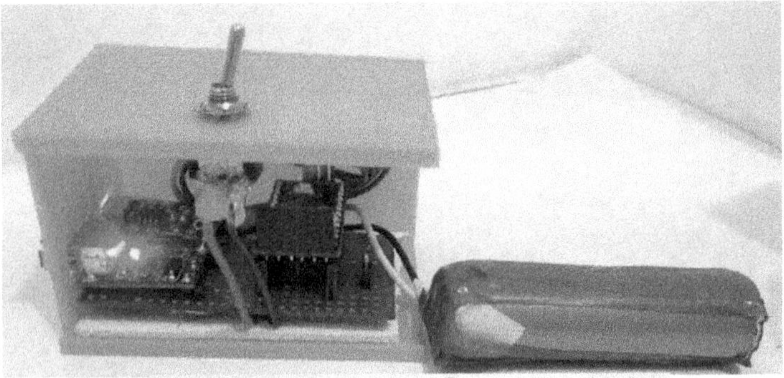

That is it our equipment is prepared for the last mounting. Just interface the corridor sensor module with the principle board and slide the mobile encase into the fixed walled in area and it's good to go.

Working Explanation:

Subsequent to ensuring your lithium battery is energized, basically turn on the pack by utilizing the flip switch and open your Android application. In the event that everything goes well you ought to get the beneath screen and it should show that your application has associated with your Bluetooth module HC-05 as demonstrated as follows. Make sure to combine your Bluetooth module with telephone before opening the application.

Presently essentially ride your vehicle and you should see the speedometer demonstrating the present speed of your vehicle. You can likewise charge you cell phone while riding by utilizing a typical charger link. When you are finished with your ride, you can slide off the case from the cycle and charge it from AC mains by utilizing any advanced cell versatile charger.

So this how, you can quantify the speed of your vehicle as well as charge the Mobile simultaneously. Expectation, you delighted in the undertaking. You can include application significantly more component to this venture, just by tweaking the codes. You can compute the separation secured by your ride, the top and the normal speed of your ride and so on. Fill me in regarding whether you have any quires through the remarks and I will be glad to enable you to out.

Code

/*Arduino Code for measuring speed of the Vehicle using Hall Sensor

*/
/*CONNECTION DETIALS
* Arduino D11 -> RX of BT Module
* Arduino D12 -> Tx of BT
* Arduino D2 -> Hall sensor 3rd pin
*/
#include <SoftwareSerial.h>// import the serial library
SoftwareSerial Cycle_BT(11, 12); // RX, TX
int ledpin=13; // led on D13 will show blink on / off
int BluetoothData; // the data given from Computer
float radius_of_wheel = 0.33; //Measure the radius of your wheel and enter it here
volatile byte rotation; // variale for interrupt fun

```
must be volatile
float timetaken,rpm,dtime;
int v;
unsigned long pevtime;


void setup()
{
  Cycle_BT.begin(9600); //start the Bluetooth com-
munication at 9600 baudrate
  //pinMode(ledpin,OUTPUT); //LED pin aoutput for
debugging
  attachInterrupt(0, magnet_detect, RISING); //sec-
ound pin of arduino used as interrupt and magnet_de-
tect will be called for each interrupt
  rotation = rpm = pevtime = 0; //Initialize all vari-
able to zero
}


void loop()
{
 /*To drop to zero if vehicle stopped*/
  if(millis()-dtime>1500) //no magnet found for
1500ms
 {
 rpm= v = 0; // make rpm and velocity as zero
 Cycle_BT.write(v);
 dtime=millis();
 }
 v = radius_of_wheel * rpm * 0.37699; //0.33 is the ra-
```

dius of the wheel in meter
}


```
void magnet_detect() //Called whenever a magnet is
detected
{
 rotation++;
 dtime=millis();
 if(rotation>=2)
 {
  timetaken = millis()-pevtime; //time in millisec for
two rotations
  rpm=(1000/timetaken)*60;   //formulae to calcu-
late rpm
  pevtime = millis();
  rotation=0;
  Cycle_BT.write(v);
  //Cycle_BT.println("Magnet detected...."); //enable
while testing the hardware
 }
}
```

10. ARDUINO RADAR SYSTEM UTILIZING PROCESSING ANDROID APP AND ULTRASONIC SENSOR

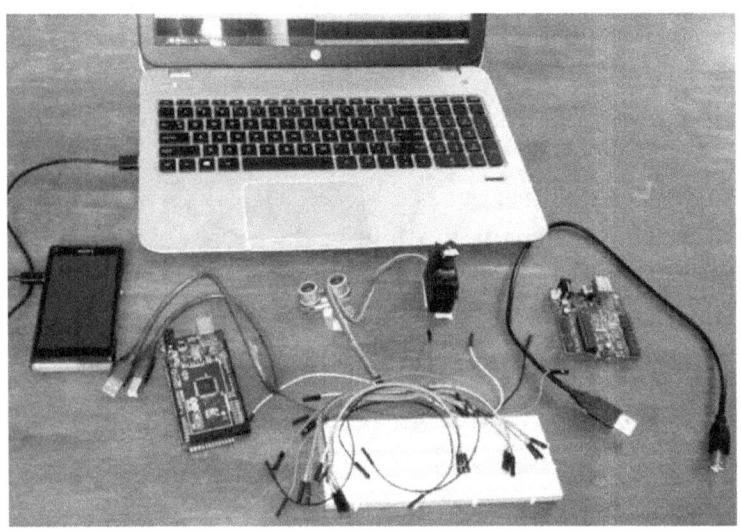

This is a fascinating undertaking with regards to which we investigate the intensity of an Arduino and Android to make a Surveillance gadget which uses Arduino and Ultra Sonic Sensor to communicate the data to a versatile application (Android) utilizing Bluetooth.

Wellbeing as well as Security has been our essential worry since ages. Introducing a surveillance camera that has night mode with tilt and dish choice will consume a major gap on our pockets. Thus let us make a monetary gadget which does nearly the equivalent yet with no video highlights.

This gadget detects objects with the assistance of Ultrasonic Sensor and subsequently can work in any event, during evenings. Additionally we are mounting the US (Ultra Sonic) sensor over a servo engine, this servo engine can be either be set to pivot consequently to check the zone or can be turned physically utilizing our Mobile application, so we can center the ultrasonic sensor in our necessary course and sense the articles present over yonder. All the data detected by the US sensor will be communicated to our Smart telephone utilizing Bluetooth Module (HC-05). So it will work like a Sonar or a Radar.

Fascinating right?? .... Give us a chance to perceive what we would require to do this task.

Requirements:

Equipment:

- A +5V power supply (I am utilizing my Arduino (another) board for power supply)

- Arduino Mega (You can utilize anything from genius smaller than expected to Yun)

- Servo Motor (any evaluating)

- Bluetooth Module (HC-05)

- Ultra Sonic Sensor (HC-SR04)

- Breadboard (not required)

- Interfacing wires

- Android portable

- PC for programming

Programming:

- Arduino Software

- Android SDK

- Handling Android (To make portable application)

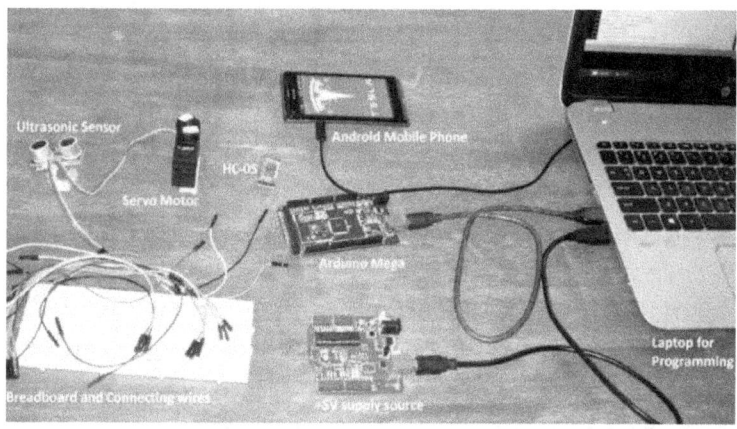

When we are prepared with our materials, let us start assembling the equipment. I have part this instructional exercise into Arduino Part and the Processing Part for simple comprehension. Individuals who are new to handling need not fear much on the grounds that the total code is given toward the finish of the instructional exercise which can be utilized thusly.

Downloading and Installing Softwares:

The Arduino IDE can be introduced from here. Download the product as per your OS and introduce it. The Arduino IDE will require a driver to speak with your Arduino Hardware. This driver ought to get introduced naturally once you interface your board with your PC. Have a go at transferring a flicker program from guides to ensure you Arduino is fully operational.

The Processing IDE can be introduced from here.

Handling is a great open source application which can be utilized for some things, it has different modes. In "Java Mode" we can make windows PC applications (.EXE records) and in "Android mode" we can make Android versatile Applications (.APK documents) it additionally has different modes like "Python mode" where you can keep in touch with you python programs. This instructional exercise won't cover the nuts and bolts of Processing, subsequently on the off chance that you need learn java programming or handling head on to this extraordinary YouTube channel here.

Arduino Hardware part and Circuit Diagram:

This task includes a ton of parts like the Servo Motor, Bluetooth Module, Ultrasonic Sensor and so forth. Henceforth in case you are an outright learner, at that point it would be prescribed to begin with some fundamental instructional exercise which includes these parts and afterward return here. Look at our different activities on Servo Motor, Bluetooth Module and Ultrasonic Sensor here.

All parts are not controlled by the Arduino itself on the grounds that, the servo engine, Bluetooth module and US sensor by and large draws a great deal of current which the Arduino won't have the option to source. Subsequently it is carefully prudent to utilize any outside +5V supply. In the event that you don't have an outer +5V supply at your compass, you can share the parts between two Arduino sheets as I have done. I have associated the Servos power rails to another Arduino board (red shading) and associated the Bluetooth module HC-05 as well as Ultrasonic sensor HC-SR04 to the Arduino mega. Alert: Powering up every one of these modules utilizing one Arduino block will sear the Arduino voltage controller.

Association outline for this Arduino Based Sonar Project is given beneath:

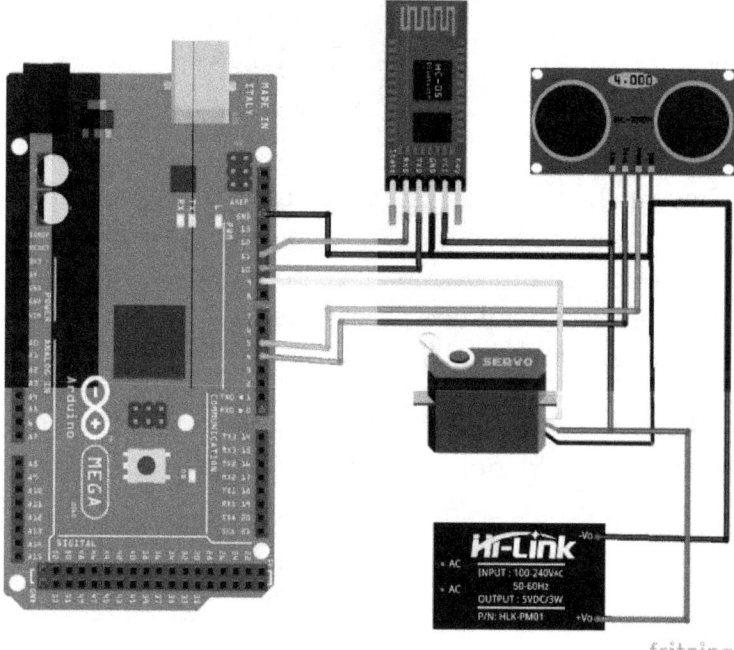

When the associations are made, mount the US sensor onto your Servo engine as demonstrated as follows:

I have utilized a little plastic piece that was in my garbage and a twofold side tape to mount the sensor. You can concoct your own plan to do likewise. There are likewise servo holders accessible in showcase which can be used for a similar reason.

When the Servo is mounted and the Connections are given, it should look something like this.

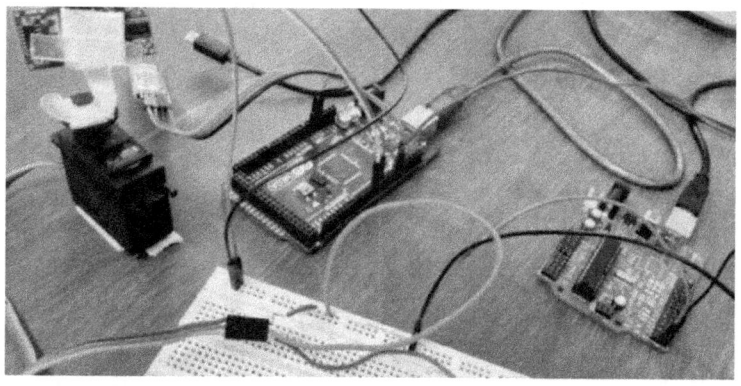

Pursue the schematics on top if get any associations wrong. Presently let us start programming the Arduino Mega utilizing the Arduino IDE.

Arduino Software Part:

We need to compose our code with the goal that we can figure the separation between an article and Ultra Sonic sensor and send it to our portable application. We likewise need to compose code for our servo engine to clear and furthermore get controlled from the information got by the Bluetooth module. Be that as it may, don't stress the program is part more straightforward than you can picture, because of Arduino and its libraries. The total code is given beneath in the code segment.

Underneath capacity is utilized to make the servo engine naturally clear from left to right (170 to 10) and again from right to left (10 to170). The two for circles are utilized to accomplish the equivalent. The capacity us() is called inside the two capacities to ascertain the separation in the middle of the sensor and the article and communicate it to the Bluetooth. A postponement of 50 ms is given to cause the servo to turn gradually. The more slow the engine pivots the precise your readings become.

```
//**Function for servo to sweep**//

void servofun()

{

  Serial.println("Sweeping"); //for debugging

  for(posc = 10;posc <= 170;posc++)  // Using 10 to
170 degree is safe than 0 to 180 because some servo
might not be operational at extreme angels

  {

    servo.write(posc); // set the position of servo
motor

    delay(50);

    us();  //measure the distance of objects sing the US
sensor
```

```
   }

   for(posc = 170;posc >= 10;posc--)

   {

      servo.write(posc);

      delay(50);

      us();   //measure the distance of objects sing the
US sensor

   }

   Serial.println ("Scan Complete"); //for debugging

   flag=0;

}

//**End of Servo sweeping function**//
```

As said before the servo engine can likewise be controlled physically from the advanced mobile phone. You essentially swipe right to make the engine move right and swipe left to make the engine move left. The above capacity is utilized to accomplish the equivalent. The heavenly attendant of the servo engine will be straightforwardly be gotten by the Bluetooth module and put away in the variable BluetoothData, at that point the servo is position in that specific holy messenger by utilizing the line servo.write(BluetoothData).

```
//**Function to control Servo manually**//

void manualservo()

{

us();

// Get value from user and control the servo

 if(Blueboy.available())

{

BluetoothData=Blueboy.read();

Serial.println(BluetoothData);

 servo.write(BluetoothData);

 Serial.println("Written");

 if(BluetoothData == 'p')
```

```
  {

    flag=0;

    }

  }

}
//__End of manual control function__//
```

The separation present before the item will be determined by beneath work. It works with a basic formulae that Speed = Distance/time. Since we know the speed of the US wave and the time taken the separation can be determined utilizing the above formulae.

```
//**Function to measure the distance**//

void us()

{

int duration, distance;

  digitalWrite(trigPin, HIGH);
```

```
  delayMicroseconds(1000);

  digitalWrite(trigPin, LOW);

  duration = pulseIn(echoPin, HIGH);

  distance = (duration/2) / 29.1; // Calculates the
distance from the sensor

  if(distance<200 && distance >0)

Blueboy.write(distance);

}

//__End of distance measuring function__//
```

Thus, when we are prepared with our code we can straight away dump the code into our equipment. In any case, the reconnaissance gadget won't begin working till it is associated with the Android Application.

Android Mobile Application for Ultrasonic Radar:

In case you would prefer not to make your own application and rather simply need to introduce a similar application utilized in this instructional exercise you can pursue the means beneath.

1. You can straightforwardly download the APK

record from the beneath connect. This APK document is made for Android variant 4.4.2 or more (Kitkat an above). Concentrate the APK record from the compress document.

Android Application for Ultrasonic Radar

2. Move the .Apk document from your PC to your cell phone.

3. Empower introducing application from Unknown sources in your android settings.

4. Introduce the application.

In the event that effectively introduced, you will discover the application named "Zelobt" introduced on your telephone as demonstrated as follows:

On the off chance that you have introduced this APK, at that point you can avoid the beneath part and hop to the following area.

Programming your own Application utilizing Processing:

It is possible that you can utilize the .APK document given above or you can manufacture your own application utilizing Processing as clarified here. With little information on programming it is likewise extremely simple to compose your very own code for your android application. Be that as it may in the event that you are simply starting, at that point it isn't fitting to begin with this code since its somewhat high than the novice level.

This program utilizes two libraries specifically, the "Ketai library" and the "ControlP5 library". The ketai library is utilized to control all the equipment present inside our cell phone. Things like you telephones battery level, nearness sensor esteems, accelerometer sensor esteems, Bluetooth control alternatives and so on can be effectively gotten to by this library. In this program we utilize this library to build up a correspondence between the telephones Bluetooth and the Arduino Bluetooth (HC-05). The "ControlP5 library" is utilized to plot diagrams for our radar framework.

The total android program is appended, you can download it from here.

Alert: Do not neglect to introduce the previously mentioned libraries and don't duplicate glue the code part alone, on the grounds that the code imports pictures from information envelope which by and large is given in above connection. Henceforth download and utilize just that.

When you are finished with the coding part and have effectively gathered it you can straightforwardly associate your cell phone to your PC through information link and snap on play catch to idiotic the application onto your cell phone. Likewise check our other Processing Projects: Ping Pong Game utilizing Arduino and Smart Phone Controlled FM Radio utilizing Processing.

Working Explanation:

Presently, we are prepared with our equipment and the product part. Catalyst your equipment and pair your versatile to the Bluetooth module. When matched open your "Zelobt" application that we just introduced and now hang tight for a second and you should see your Bluetooth module (HC-05) naturally getting associated with your advanced mobile phone. When the association is built up you will get the accompanying screen:

You can see that it says associated with: Device name (equipment address) on the highest point of the screen. It likewise shows the present heavenly attendant of the servo engine and the separation between the US sensor. A blue diagram is additionally plotted on the red foundation dependent on the deliberate separation. The closer the article gets the more the taller the blue region gets. The chart estimated when a few articles are set close is additionally appeared in the second figure above.

As said before you can likewise control your servo engine from your versatile application. To do these, basically click on stop button. This will prevent your servo from clearing consequently. You can likewise locate a roundabout wheel at the base of the screen which when swiped will pivot in clock or hostile to clock astute bearing. By swiping this wheel you can likewise make your servo engine turn in that specific heading. The haggle diagram refreshed when swiped are appeared in the image underneath.

Arduino Code is given underneath and the APK document for android Application is here. Expectation you comprehended the undertaking.

Code

```
/*
* DIY-RADAR
*
```

```
* US Sensor
* trigpin >> 4
* echopin >> 5
*
* Servo motor
* signal pin >> 9
*
* Hc-05
* tx >> 10
* Rx >> 11
*
* WARNING
* Do not power up all your modules to a single Ar-
duino the current drawn will fry your arduino board.
Use separate supply.
*
*/
//*Include the required header files**//
#include <Servo.h>
#include <SoftwareSerial.h>
//__End of including headers__//
//**Defining pins for US sensor**//
#define trigPin 4
#define echoPin 5
///__End of defaniton__//

SoftwareSerial Blueboy(10, 11); //Naming our Blue-
tooth module as Blueboy and defiing the RX and TX
pins as 10 and 11
Servo servo; //Initializing a servo object called servo
//**Global variabel declarations**//
```

```
int BluetoothData;
int posc = 0;
int flag=10;
//__End of global variable declartion__//


void setup() //Runce only once
{
  servo.attach(9); //Servo is connected to pin 9
  pinMode(trigPin, OUTPUT); //trigpin of US sensor is
output
  pinMode(echoPin, INPUT); //echopin of US sensor is
Input
   Serial.begin(38400); //Serial monitor is started at
38400 baud rate for debugging
  Blueboy.begin(9600); //Bluetooth module works at
9600 baudrate
   Blueboy.println("Blueboy is active"); //Conforma-
tion from Bluetooth
}
void loop() //The infinite loop
{
//**Program to start or stop the Survilance devide**//
  if(Blueboy.available())
{
Serial.println("Incoming"); //for debugging
BluetoothData=Blueboy.read(); //read data from
bluetooth
Serial.println(BluetoothData); //for debugging
  if(BluetoothData == 'p') //if the mobile app has sent
```

```
a 'p'
 {
 flag=0; //play the device in auto mode
 }
 if(BluetoothData == 's') //if the mobile app has sent a
's'
 {
 flag=1; //stop the device and enter manual mode
 }
Serial.println(flag); //for debugging
}
if(flag==0)
servofun(); //Servo sweeps on own
if(flag==1)
manualservo(); //Manual sweeping
}
//_End of loop program__//
//**Function for servo to sweep**//
void servofun()
{
 Serial.println("Sweeping"); //for debugging
 for(posc = 10;posc <= 170;posc++) //Using 10 to 170
degree is safe than 0 to 180 because some servo might
not be operational at extreme angels
 {
 servo.write(posc); // set the position of servo motor
 delay(50);
 us();   //measure the distance of objects sing the US
sensor
 }
```

```
  for(posc = 170;posc >= 10;posc--)
  {
   servo.write(posc);
   delay(50);
   us();   //measure the distance of objects sing the US
sensor
  }
  Serial.println ("Scan Complete"); //for debugging
  flag=0;
}
//**End of Servo sweeping function**//
//**Function to control Servo manually**//
void manualservo()
{
us();
// Get value from user and control the servo
  if(Blueboy.available())
{
BluetoothData=Blueboy.read();
Serial.println(BluetoothData);
  servo.write(BluetoothData);
  Serial.println("Written");
  if(BluetoothData == 'p')
  {
  flag=0;
  }
}
}
```

```
//__End of manual control function__//
//**Function to measure the distance**//
void us()
{
int duration, distance;
 digitalWrite(trigPin, HIGH);
 delayMicroseconds(1000);
 digitalWrite(trigPin, LOW);
 duration = pulseIn(echoPin, HIGH);
 distance = (duration/2) / 29.1; // Calculates the dis-
tance from the sensor
 if (distance<200 && distance >0)
Blueboy.write(distance);
}
//__End of distance measuring function__//
```

Thank you !!!